HIDDEN GOLD

THE CENTRAL NORSEMAN STORY

MONOGRAPH SERIES

1.	* Detrital Heavy Minerals in Natural Accumulates	*George Baker*	*1962*
2.	Research in Chemical and Extraction Metallurgy	*Ed. J. T. Woodcock, A.E. Jenkins and G. M. Willis*	*1967*
3.	* Broken Hill Mines – 1968	*Ed. M. Radmanovich and J. T. Woodcock*	*1968*
4.	* Economic Geology of New Zealand	*Ed. G. J. Williams*	*1974*
5.	Economic Geology of Australia and Papua New Guinea – 1. Metals	*Ed. C. L. Knight*	*1975*
6.	* Economic Geology of Australia and Papua New Guinea – 2. Coal	*Ed. D. M. Traves and D. King*	*1975*
7.	* Economic Geology of Australia and Papua New Guinea – 3. Petroleum	*Ed. R. B. Leslie, H. J. Evans and C. L. Knight*	*1976*
8.	* Economic Geology of Australia and Papua New Guinea – 4. Industrial Minerals and Rocks	*Ed. C. L. Knight*	*1976*
9.	Field Geologists' Manual (Third Edition)	*Ed. D. A. Berkman*	*1989*
10.	* Mining and Metallurgical Practices in Australasia (The Sir Maurice Mawby Memorial Volume)	*Ed. J.T. Woodcock*	*1980*
11.	Victoria's Brown Coal – A Huge Fortune in Chancery (The Sir Willis Connolly Memorial Volume)	*Ed. J.T. Woodcock*	*1984*
12.	Australasian Coal Mining Practice	*Ed. C.H. Martin*	*1986*
13.	Mineral Deposits of New Zealand	*Ed. D. Kear*	*1989*
14.	Geology of the Mineral Deposits of Australia and Papua New Guinea	*Ed. F. Hughes*	*1990*
15.	The Rocks Speak	*H. F. King*	*1989*
16.	Hidden Gold – The Central Norseman Story	*J. D. Campbell*	*1990*

Key: * = Out of Print

Copies of all books currently in print can be obtained from
The Australasian Institute of Mining and Metallurgy
Clunies Ross House 191 Royal Parade Parkville Victoria Australia 3052
Telephone (03) 347 3166; Facsimile (03) 347 8525.

HIDDEN GOLD

THE CENTRAL NORSEMAN STORY

An account of Structural Geological Studies and Ore-search
at Norseman, Western Australia

J. D. CAMPBELL

VOLUME ONE
accompanied by volume two
figures and plates

INCORPORATED BY ROYAL CHARTER 1955

MONOGRAPH 16

Published by
THE AUSTRALASIAN INSTITUTE OF MINING AND METALLURGY
CLUNIES ROSS HOUSE 191 ROYAL PARADE PARKVILLE VICTORIA AUSTRALIA 3052

The generous support of Western Mining Corporation Holdings Limited and
Central Norseman Gold Corporation Limited is gratefully acknowledged

Volume 1
ISBN 0 949106 45 3

Cover photograph

A Central Norseman gold quartz sample.

First published in 1990
Typeset in Century Old Style 11pt by
Print Concepts

Designed and printed in Australia by
James Squire Color Printing
76 Chestnut Street Richmond Victoria Australia 3121

In memory of a wise and kindly gentleman

Herbert St. John Summers, D.Sc.

Associate-Professor of Geology, 1936 – 1941
Professor of Geology, 1941 – 1944
University of Melbourne

to whom I owe so much.

J. D. CAMPBELL

Perth, W.A., July 1990

FOREWORD

Within my experience this is a unique and valuable publication which should be read by all practicing exploration geoscientists and should be prescribed reading for all undergraduates who aspire to become explorationists. It is not easy to read as it requires close attention to follow through the various stages of testing and development but the effort will be rewarded by the understanding of the success of the Central Norseman Story. This includes not only the detail of the geometry of the reefs and the oreshoots but an appreciation of the persistence of exploration effort necessary to yield the success which has maintained production for more than 50 years.

It is unique because very few major ore deposits have had such sustained and competent attention from an integrated group of geologists. Most major deposits have been explored in a series of disconnected campaigns frequently staffed by teams with little knowledge of previous exploration of these deposits. Past knowledge is so often lost. The Central Norseman Story has the added advantage that the author, Don Campbell, was involved with the exploration from the beginning to the present by his long association with Western Mining Corporation.

The geologists involved in this exploration and development saga were influenced from the start by the idea that if a standard of excellence in geological endeavour is maintained it will support a successful mining industry. It was over this period that mining geology was developed as an accepted and separate profession in Australia. The participants in the Central Norseman Story were in fact largely responsible for the development of mineral exploration and mining geology in Australia and have seen it accepted by the mining engineering profession which originally questioned the value of geology to the industry.

The publication also preserves some valuable history in recording the work of the various geologists who contributed to the Central Norseman Story. In particular it redirects attention to the classic paper by Terence Conolly in Economic Geology in 1936 when he described his subtle contour method of revealing vein structure. Although one would like to see similar publications on other major Australian ore deposits which have had long histories of exploration and production they are unlikely to appear because the same unusual circumstances of continuity of personal involvement do not exist. Much of the value of the Central Norseman Story lies in the Volume of Figures and Plates where the evolution of ideas has been so well preserved and presented.

It is not expected that executives of the mining industry will devote enough time and effort to read the publication in detail, but it is to be hoped that many will browse through the volumes and absorb the important message conveyed, that the persistent pursuit of excellence in exploration will be rewarded.

The Central Norseman Story is an outstanding record of the approach which is needed and the mining industry should be grateful to Western Mining Corporation for undertaking to have this experience preserved. The mineral industry in general and the geological profession in particular owe a debt of gratitude to Don Campbell for this magnificent work in which he has recorded not only the Central Norseman Story but something of himself.

Eric Rudd
Emeritus Professor
University of Adelaide

AUTHOR'S PREFACE

I started a course in mining engineering at the University of Melbourne in 1928 and this led to an interest in geology. As the mining course of those days covered nearly the whole of under-graduate geology, I decided to do the full third-year of geology even though this meant doing palaeontology, my least favourite subject. In due course, I graduated B.Sc., B.M.E.

My interest in geology derived from Dr. H.S. Summers, to whom this volume is dedicated. "Doc.", as he was known to all and sundry including the under-graduates, was a very "human" man with a great sense of fun who understood young people and so was at ease with them and amongst them. As he knew his subject and in fact loved it, he enjoyed imparting his knowledge to the up-and-coming generation; in other words, he was that rarest of things, a first-class and stimulating lecturer. The result was that I became greatly interested in geology, a subject that was a bit of a "chore" in their second, third and fourth years for most mining students. Doc Summers became a friend and helper during my later years at University and for many years thereafter. It was through this friendship that I met my wife – his daughter and Ruth Tonkin, as she was then, had been at school together at Merton Hall (Melbourne Church of England Girls Grammar School).

As it was more convenient to complete geology first, the study of mining engineering fell into my last year at university. The lecturer was Lindesay Clark and this led, early in 1933, to his offering me a position with the about-to-be-formed Western Mining Corporation. I was the first staff member on site and, for the next six weeks or so, was the only one – one "Chief" and one "Indian", as it were.

I intended to take up mining but, towards the end of 1933, was asked by Lindesay to go to Norseman to assist Conolly, one of W.M.C.'s recent imports from North America. It was hoped that I would learn from him something of the relatively new field of structural geology. I was a reluctant starter but, after a few weeks at Norseman, became a willing and enthusiastic convert. This enthusiasm has never dwindled and Norseman has remained an abiding interest.

I have felt for many years that the story of Norseman should be recorded but the idea did not crystallise until October, 1984. I had been invited to the opening of the new gold mining operation at Stawell in Victoria, an area tested intermittently by W.M.C. from 1944 until the 1960's but not viable until recent years. On the way home, I mentioned the Norseman idea to Roy Woodall. He felt as I did and arranged for the project to go ahead with support from Western Mining and Central Norseman. We had in mind a time span of a year or two but, in the event, this estimate proved a little optimistic.

From its earliest days, Western Mining has used a system of K reports to record significant recommendations and analyses (K derives from Kalgoorlie). My time estimate assumed that those relating to Norseman needed only to be digested, co-ordinated and rounded-out to give an account of the history of the operation. I was wrong. In the main, these reports were written on specific subjects for specific purposes and assumed, with few exceptions, a good knowledge of the Norseman operation. In consequence, their illustrations and contents did not give the over-all picture needed by readers unfamiliar with Norseman.

I found myself back therefore to my early days with Conolly, as the need was to make complete interpretations so that the ideas developed by many people over many years could be related to their geological settings. This proved to be a large task but a pleasant one. The requirement for doing detailed interpretation is a large amount of un-interrupted time so that ideas can be pursued until the best answer is found. Finality has taken a long time but it has been an interesting journey.

The sixty-one figures and twenty coloured plates illustrating the Central Norseman Story are derived from this study and from the diagrams accompanying the many K reports. In themselves, they tell the story but they do need description and connective tissue. The text gives background, explanations of the illustrations and the geological concepts under-pinning them and deriving from them. There was a progression of ideas at Norseman, a few of them disastrous, and all are recorded in the illustrations and the text. Success came from analysing the failures, particularly in the earlier stages of the operation, but it was a slow process. It took ten years to yield the first dividend and it was another ten years before the company was really prosperous. The benefits extended beyond Norseman as the revenue deriving from Central Norseman during and after the 1950's was a key factor in the rise to prominence of its major shareholder, Western Mining Corporation.

CONTENTS

FIRST OF TWO VOLUMES

Foreword	vii
Author's Preface	ix
INTRODUCTION	1
GENERAL GEOLOGY	5
FRACTURE PATTERNS	7
MARAROA REEF	9
Initial Studies and Exploration	9
Vein Contouring	12
The Ajax Shaft	16
The Favourable Bed Hypothesis	17
Diffusion of Fractures	19
Fracture Patterns, S Structures	20
The Folding Hypothesis	23
PRINCESS ROYAL – NORTH ROYAL OREBODY	25
Bonanza	27
Shear Link Pattern	30
Near-Surface Ore	31
The S Structure	32
The Open-Cuts	34
General Structure	35
PRINCESS ROYAL – DEEP & NORTHERN EXPLORATION	39
Exploration in Depth	39
Northern Exploration	41
NORSEMAN REEF	43
CROWN REEF	46
Early Mining and Testing	46
Development of Crown Reef	47
Reef Patterns and Ore Shoots	48
Mining	50
THE BLUE BIRD SEQUENCE	51
EAST-DIPPING DYKES	54
ORE PATTERNS	55
PRODUCTION HISTORY	60
IN RETROSPECT	63
REFERENCES	68
FIGURES AND PLATES	Volume 2

INTRODUCTION

The Norseman gold field, about 150 kms south of Kalgoorlie, was found in 1894 following minor discoveries two years earlier at Dundas, 22 km to the south. The quartz reefs of the Norseman area, although narrow, have considerable length and account for nearly all the production of the Dundas Goldfield – 10,605,000 tonnes of ore for the production of 4,433,600 ounces of gold, an average grade of 13.0 gms. per tonne. The main contributors have been the Mararoa, Princess Royal and Crown reefs, with lesser but significant productions from the Norseman and Cumberland reefs. Other reefs of varying consequence have added to the total.

Productions for the more significant periods in the history of the Dundas Goldfield are given below.

Years	Tonnes	Ounces	g/t
1894 – 1923	921,000	617,800	20.9
1924 – 1932	33,000	41,200	38.8
1933 – 1952	2,836,000	911,800	10.0
1953 – 1987	6,815,000	2,862,800	13.1
Total	10,605,000	4,433,600	13.0

As these figures indicate, the field was moribund from 1924 to 1932 due partly to the low price of gold (4.25 Pounds per oz.) in relation to rising costs but mainly to the fact that the orebodies of the original discoveries had been worked out or were approaching exhaustion.

Interest in the field revived at the beginning of the 1930s following the English devaluation of Sterling and the subsequent decision of the United States to allow gold to rise from its fixed value of $20 per ounce to a maximum of $35 per ounce. It stood for most of the pre-war years at $32 per ounce, generally equivalent to 9.5 Pounds in Australian currency.

Recorded productions from the reefs at Norseman to the end of 1933 are given below:

Reef	Tonnes	Ounces	g/t
O.K.	11,000	15,320	43.3
Cumberland	67,700	65,120	30.1
Princess Royal	183,300	152,150	25.8
Norseman	158,900	107,620	21.1
Mararoa	352,200	171,660	15.2
Other Reefs	113,700	67,000	18.3
Total	886,800	578,900	20.3

The Mararoa reef was the biggest producer but at a grade lower than those of the other reefs. However, it was the focus of attention in the early 1930s, due to three factors:

1. Rising costs had forced the old Mararoa company to close in 1924 leaving modest ore occurrences going under-foot in various parts of the mine;

2. Immediately to the south of the Mararoa mine, Norseman Gold Mines N.L. was developing the recently discovered Butterfly orebody on the Mararoa reef;
3. The other major orebodies, although more attractive in grade, had been worked-out in the earlier days of the field.

In 1933, when Western Mining Corporation sought to enter the field, Norseman Gold Mines held the southern end of the Mararoa reef and the old Mararoa company workings were held by Phoenix Gold Mines N.L., a small Adelaide-based company. Phoenix Gold Mines agreed however to grant Western Mining an option of purchase, as did some prospectors who held the northern end of the Mararoa out-crop. The deeper reaches of the reef, which dips east at about 45°, were pegged by Western Mining giving it control of a considerable area if exploration should find significant orebodies.

The search for ore was guided by methods of structural geology which had been developed in the early years of this century at Harvard University. They had been applied successfully at the Homestake gold mine in South Dakota by Dr. Donald McLaughlin in the 1920s and soon afterwards by Dr. Hugh McKinstry in the Porcupine goldfield at Timmins in Canada. A fortunate circumstance was that Dr. McLaughlin was a consultant to Case Pomeroy, an American member of the group of companies that founded Western Mining. On his recommendation, Dr. McKinstry led a group of young American and Canadian geologists who came to Australia during 1933 to introduce this new approach into the operations of Western Mining Corporation*.

This method used careful and detailed mapping followed by complete interpretation, three-dimensional where possible. A key point was a clear indication of the factual basis on the final presentation, the interpreted plans and cross sections. It proved enlightening at Norseman and suggested that there was real promise not far beyond the limits reached by the old Mararoa company. This was found to be true but the search for new orebodies on the Mararoa reef proved arduous and more difficult than had been expected.

The problem was not the concepts but their limitations. Study of the oreshoots and vein pattern of the old Mararoa workings showed that the system should continue in depth and also longitudinally but this analysis gave no specific guides. Exploration had therefore to be a learning process. It was directed to the areas thought to be the most likely and had to be redirected from time to time as new information came in and gave a better understanding of the geology and the exploration problem.

A difficulty was the overall pitch of the ore system. It was evident from the outset that the individual orebodies of all the reefs at Norseman pitch south at a moderate angle. It was apparent also on the Mararoa reef and particularly on the Norseman reef, which is parallel and nearby, that there is a near-horizontal influence as well. As work progressed, the data favoured first the one pitch direction and then the other as being the more

* *The two "Canadians" were H. J. C. ("Terence") Conolly, of Anglo-Irish parentage, born in India and educated in England, and Haddon King, born in British Guiana. After working together in British Guiana, they had moved to Canada in the 1920s. Haddon received his tertiary education at the University of Toronto. Both were to play significant roles in the earlier years of the Norseman operation.*

important. The problem was not resolved until 1941 when a disastrous result in the Ajax Shaft, sunk to command the deeper reaches of the Mararoa reef, showed that the horizontal influence is the main factor and that the individual orebodies pitch south within a zone trending horizontally or gently north.

This concept was the turning point at Norseman. It gave direction, previously lacking, to the development and exploration of the Mararoa reef and its application farther afield led to the finding of the major prize of the district, the Princess Royal North orebody. Another success, some years later, was on the Crown reef on the western side of the Mararoa workings. A recent result has been the finding of a small orebody at the southern end of the Mararoa reef.

Central Norseman Gold Corporation N.L. was formed in 1935 to take over the Western Mining Corporation interest in the area. Production was commenced during 1936 on ore which proved to be low grade. However, things improved several years later.

The production history of the company is given below:

Years*	Tonnes	Ounces	g/t
1937 – 1939	161,700	37,400	7.2
1940 – 1943	403,900	169,200	13.0
1944 – 1952	1,027,500	335,200	10.1
1953 – 1969	3,083,000	1,619,000	16.3
1970 – 1975	857,500	261,500	9.5
1976 – 1988	2,874,900	1,028,880	11.1
1937 – 1988	8,408,500	3,451,100	12.8

Average tonnes mined per year for these periods were:

1937-39	1940-43	1944-52	1953-69	1970-75	1976-88
54,000	101,000	114,000	179,000	143,000	221,000

The first profitable years were 1940-43 and the first dividend was paid in 1943, ten years after entry into the field and seven years after commencing production. The ore came from the Mararoa reef, mainly from the newly developed No.3 Orebody.

Profits declined after 1943 due to falling production caused by war-time difficulties, improved slowly after 1947 as more labour became available, and then rose three-fold or more when the new North Royal orebody came into full production in the year ending March 1953. A successful operation had become very profitable.

Grade declined during the period 1970-75 as underground production from North Royal dwindled and died. The first loss since 1939 was made in 1971 and the cessation of the Norseman operation was considered. Fortunately the price of gold commenced its improvement before closure became inevitable.

* *Year ending March until 1965; subsequently, year ending June.*

Soon afterwards, improved drilling techniques allowed the first effective testing of soft and deeply oxidised ground above the North Royal orebody. This led to the discovery of extensive high-grade ore reaching almost to the surface and, several years later, to a significant improvement in Central Norseman's production grade. The problem at Princess Royal of mining in very soft ground was met by using open-cut methods.

A feature at Norseman is that, although all ore occurrence is relatively shallow, most of it did not reach the surface. The total production to date for the field is 10,605,000 tonnes for 4,433,600 ounces at 13.0 g/t of which about 500,000 tonnes of a grade of some 23 g/t derive from the discoveries of the early days. Most of the production has come from non-outcropping (or "blind") orebodies.

GENERAL GEOLOGY

The general geology of the Norseman area is shown in *Figure 1.* The area of interest lies between Lake Cowan on the west and Lake Dundas on the east. These salt lakes are old river valleys now deeply filled with Tertiary marine sediments and more recent wind-drifted material.

The oldest rocks of the area, Archaean in age, form a thick sequence of sandstones, quartzites, mudstones and conglomerates with some interlayered basaltic lavas and intrusive dolerites, mainly sills. These are called the Noganyer Group. The rocks of this group dip to the west at about 50° and are overlain conformably by a great thickness of Archaean basaltic rocks, mainly sub-marine lava flows with some intrusive dolerites and gabbros and occasional thin bands of sediments between lava flows.

To the north of Norseman, this sequence is interrupted by an east-west trending, vertically dipping norite dyke one to two kilometres wide and also of Archaean age. It is hundreds of kilometres long and is a major geological feature of great interest to petrologists. However, it is of no consequence economically as there is no significant movement along it and the Norseman rock sequence continues on its northern side. Noganyer sediments are found on the northern side of the norite dyke in a position a little to the east of the area shown in *Figure 1.*

The quartz reefs of Norseman lie in the basaltic rocks, trend north-south and dip to the east at about 40° – 45°. The main reefs of the area are the Mararoa and Norseman, with the Maloney and Crown to the west of the Mararoa and the Valkyrie, Lady Jean and Mt. Barker reefs to the east of the Norseman. As the name implies, the Norseman reef was the initial discovery in the area and its southern end was the site of an early major operation named, appropriately, Norseman Gold Mines.

Another major operation of the early days was to the north of the norite dyke where the Princess Royal reef outcrops at the southern tip of flat-lying Miocene sediments. These young sediments are of small significance geologically but of great importance economically.

In Miocene times, the climate was mild and moist and Lake Cowan was the site of a major river with its bed 100 metres or so below the present salt lake surface. It led northwards to the neighbouring Lake Lefroy and may have extended towards Kalgoorlie and perhaps farther. To the south, it drained towards Esperance.

During the Miocene period, there was a general submergence of the Australian continent and an arm of the sea reached up this river valley to and beyond Norseman. The valley was filled with marine sediments consisting of clays, limestones and spongolite beds.[(1)(2)] Subsequent elevation to the present position and the dry climate following the Pleistocene glaciation led to this valley and others becoming chains of salt lakes.

The point of interest is that the Archaean rocks enclosing the Princess Royal reef are very soft and easily eroded. In consequence, in the Miocene period, the outcrop line was the site of a side valley draining into the Lake Cowan river of those times. It was filled by Miocene sediments. This fortuitous circumstance caused an almost complete concealment of the reef – to the great benefit of Central Norseman, once the northern continuation of the reef had been found.

Figure 2 is a sketch cross-section along the line A-A of *Figure 1* – at 125ON co-ordinate, the latitude of the Phoenix Shaft. The Norseman area is on the west limb of an anticline

1. *W. D. Campbell, G.S.W.A., Bulletin No. 21, 1906.*
2. *E. de C. Clarke, C. Teichert, J. R. H. Mc Whae, Tertiary Deposits near Norseman, W.A., Proc. Roy. Soc. of W.A., 1948.*

which lies somewhere to the east of the cross-section. The location of the syncline to the west is also unknown. The basaltic sequence extends at least to the eastern shore of Lake Cowan, giving it a thickness of 3 kilometres and probably more, perhaps much more.

The only indications of fold structure are found in the Noganyer Group to the east of the mines. This folding is seen best in inter-calated fine-grained ferruginous sediments, seen at the surface as banded iron formations or B.I.F.'s. They show minor folding which is west limb of anticline pitching north. Opposite the mines at Norseman, they have pitches ranging from nearly horizontal to 40° north; towards the norite dyke, they steepen to 45° – 60° north and reach 90° in some places. Some kilometres to the south of Norseman, these minor folds turn to a south pitch and become steeply south at Dundas, about 20 kilometres south of Norseman. The orebodies of Norseman lie therefore on the northern side of a strong "cross-fold".

FRACTURE PATTERNS

Although the general geology given in *Figure 1* and the reef system presented in *Figure 2* seem simple, they both have their origins in a major disturbance of the earth's crust. The originally flat-lying sequence of sediments and volcanics was strongly folded in Archaean times and the rocks seen today at the surface would have lain then at a considerable depth within a major belt of mountains, not dissimilar probably to the present-day Himalayas or Andes.

The reaction to stress of deeply buried rocks is unlike their behaviour at the surface as they have to support the weight of the over-lying material. At a depth of 5 miles, the resulting pressure is about 30,000 lbs. per sq. inch, equal to the crushing strength of the strongest rocks under surface conditions. This pressure is exerted in all directions, i.e. it is a confining pressure, and so the rocks do not crush. Instead, although remaining strong, they tend towards plastic behaviour.

This concept led early in this century to the recognition that the strain ellipsoid, used by engineers and scientists to explain the behaviour of ductile materials, could be applied to the stress-strain relationships occurring in geological structures. The basic elements as they are thought to apply at Norseman are presented in *Figure 3.*

Figure 3A shows the side of a cube of rock of dimensions *a* and *b*, *a* being equal to *b* of course. Under an east-west compression, *a* contracts to *a'* and *b* expands to *b'*, *a* x *b* being approximately equal to *a'* x *b'*, *Figure 3B.* This figure shows the point at which the elastic limit has been reached and shows also two potential shearing planes. These dip approximately 45° east and 45° west and each will yield a reverse fault, i.e., for the east-dipping fracture, the movement will be east-block-up and, for the west-dipping fracture, west-block-up, as is indicated by the arrows.

There is also a secondary factor. The upwards stretching of the cube of rock will add a tension effect to the stress system which, on its own, would produce horizontal tension cracks as is sketched in *Figure 3C.*

These effects were demonstrated experimentally by Hans Cloos[3] at the beginning of the 1930s. He used slabs of stiff modelling clay as these were thought to correspond in their general characteristics to those of deeply-buried rocks.

Figure 3D sketches the results of a compression experiment, presented photographically by Cloos. The ellipse is the end result of the deformation of a circle stamped onto the clay at the beginning of the test. Two sets of shears have developed but fractures showing significant movement are not numerous and dip predominantly to the right. The test-piece has been markedly deformed and it is evident therefore that most of the deformation has been at the near-microscopic level (rock-flowage in geological terms). As the ellipse is not skewed, it is evident that the total movement (flowage plus visible fracturing) is more or less equal for the two directions of shearing.

Figure 3F sketches at a macroscopic scale the effect of the west-dipping flowage on the east-dipping shears. Each of the west-dipping shears in 3F has a small west block up movement, giving a "steps and stairs" pattern to the east-dipping fracture (shown as an interrupted heavy line). In actuality, of course, the west-dipping shears are very closely spaced and their combined effect would be to rotate the original shear A-A from 45° to the steeper dip B-B. The rotated shear would be continuous, as in *Figure 3D.*

In another experiment, Cloos placed a slab of clay under tension and produced

3. *H. Cloos, Zur Experimentellen Tektonik I and II, Die Naturwissenschaften, Vol. 18, No. 34, 1930; Vol. 19, No. 11, 1931.*

rudimentary tension gashes, as is sketched in *Figure 3E.* If Cloos' photographic reproduction is studied closely, it is evident that complementary shearing movements have occurred, each inclined at about 45° to the tension force. They are shown diagrammatically in *Figure 3E.* Their attitudes and directions of movement are similar to those of the compression experiment and they would in fact have contributed more to the total deformation than did the tension gashes. In other words, the two are inter-related and tension fracturing is always present, or potentially present, in any shearing deformation.

In the deformation giving rise to the reefs at Norseman, the west-dipping component of the fracture pattern was parallel to the stratification of the basaltic lavas and so was absorbed, in the main, as rock flowage. The east-dipping component was at right-angles to stratification and so had an opportunity to develop finite fractures in addition to rock flowage. The more persistent fractures provided channels for the development of quartz reefs such as the Mararoa, Norseman, etc.

In detail, these reef fractures are complex and show an interplay between shearing and tension fracturing. However, the development of these two elements is not random, as in the experiments by Cloos. Instead, they combine to form a single channel consisting of over-lapping shears joined one to the other in an orderly and progressive manner by flatter-dipping fractures called, for obvious reasons, links. The general characteristics are indicated in *Figure 3G.* Most of the links are not true tension gashes but are flatter-dipping sections of the reefs combining elements of both shear and tension. Their dips are mostly in the range of 20° east to 35° east. However, the reefs do follow true tension fractures in places and are then more or less horizontal in attitude but irregular and erratic. Such fractures can be extensive, as for instance at Princess Royal. (Note: The links are shown at exaggerated widths to give clarity of presentation. The shears are extended as dashed lines above and below their intersections with the links. Again the purpose is clarity and the dashed lines do not represent continuations of the shears. This method is used in all illustrations showing the shear-link patterns of the various mines).

MARAROA REEF

The Mararoa reef outcrops at its northern end and was found and worked in the early days of the field. It was developed by the All Nations company and to its north by the Pearl company. Neither was successful as, although they found good widths of quartz, the gold content was meagre.

To the south of the All Nations, the reef outcrop is poor and is covered by soil and hill-wash. In the early 1900s, auriferous quartz was found in this area by the Mararoa company but they had a difficult time in their early years. They sank an inclined shaft and found a short run of good ore on their No.1 level, a little more on No.2 level and slightly more on the No.3 level. Their greatest length of semi-continuous ore was some 200 feet on No.3 level. Then, to their pleasant surprise, the No.4 level really opened out – it showed continuous high-grade ore of a good width over a length of 1000 feet. This ore was on the foot-wall reef which had shown little ore on the higher levels. It proved in fact to be the main reef. The No.5 and No.6 levels were even better but the ore grade and the length of ore declined on No.7 and No.8 levels.

The Mararoa company prospered and went on to become the largest producer of the field. However, by 1918, they had mined the best of their orebody and were faced with rising costs and a declining grade. They struggled on but were forced to close in 1924.

The No.4 level of the Mararoa mine is 75m vertically below the surface. To the south of the Mararoa, the equally large Butterfly orebody comes even closer to the surface but was not found until the late 1920s and was not worked until the early 1930s – by Norseman Gold Mines. It is a characteristic of the Norseman district that the orebodies seldom reach the surface, although all occur at relatively shallow depths.

Initial Studies and Exploration

When Hugh McKinstry first examined the Norseman area in 1933, he recognised immediately the key features of the reef patterns. Also, he was impressed by the possibilities of the district. In consequence, he recommended that Terence Conolly's first task in Western Australia should be a study of the Mararoa reef at Norseman. Conolly approached the problem by two methods – firstly, by detailed mapping and interpretation of the vein structure and then by a study of the distribution of assay values.

The assay study was the simpler task and was completed first. To the surprise of Conolly, and even more to that of McKinstry, it showed the dominant trend of ore occurrence to be horizontal or nearly so. The results are presented in *Plate 1A.* *

The top of ore occurrence is just above No.4 level over most of the length of the main orebody (No.2a Orebody) and is also nearly horizontal for the full length of its northern extension (No.2b Orebody).** This is confirmed by the attitudes of most of the higher grade shoots – coloured red in Plate 1A. The northern half of the No.2a Orebody was

* *Imperial co-ordinates have been used for the Mararoa reef, i.e., co-ordinate distances are stated in feet. Unfortunately, the Norseman mine plans are all drawn looking from east to west; the normal convention is to view from the west.*

** *The orebodies along the Mararoa reef have been numbered from south to north, rather than in order of discovery. Thus the Butterfly orebody is No.1, the old Mararoa company position is No.2 and the later discoveries by Central Norseman are No.3 and 4. For convenience, the original and main position worked by the Mararoa company is described as the No.2a Orebody and its northern extension is called the No.2b Orebody.*

thought by Conolly to contain a significant feature as it showed two horizontally-disposed high-grade shoots occurring one below the other.

Mapping the mine and the subsequent interpretation of the data was a large task. The mine levels, all of which were accessible, were mapped in detail and cross-sections at 50 ft intervals were surveyed and mapped throughout the length of the mine.

The mine mapping showed that the Mararoa reef occurs within a band of actinolite schist about 50 to 100 ft wide, dipping easterly and more or less parallel to the reef. As the enclosing fine-grained basaltic rocks dip west, it was concluded that the actinolite schist was the result of shearing and hydrothermal activity along the reef-channel zone prior to final fracturing and the deposition of quartz. Conolly attempted detailed mapping of the wall-rocks of the mine openings but gave up at an early stage as it seemed unlikely to yield a useful result.

The geological work became therefore a careful mapping of the quartz reefs followed by a shear-link study of the vein system. This interpretation was done in great detail in both plan and cross section – a considerable undertaking as there were eight levels of lengths up to 2500 ft and about 50 cross-sections to be covered. Conolly's energy and drive are indicated by the fact that the mapping and interpretation were done in less than three months. He arrived in Norseman early in January 1934 and the work got under way towards the middle of the month. He presented his report soon after the end of March.[4]

Unfortunately, the plans and cross-sections carrying his interpretation have not survived. All that remains is a copy of the main plan accompanying his report. This has been used in preparing Plates 1A, 1B and 1C. To cover the loss of Conolly's original work, a re-interpretation has been done by the writer. It covers a much larger area than was available to Conolly but is done in less detail than his study.

Plate 1B shows Conolly's mapping and interpretation of the vein system superimposed on the relevant portion of the shear-link interpretation done recently. Six cross-sections are given in Figure 4. Shears are identified by letters and links by numbers.

As exposure of the reef in the northern part of the mine is limited, the main interest lies in the vein pattern of the No.2a Orebody. A feature is the occurrence of hanging-wall and foot-wall veins, as can be seen on several of the cross-sections in *Figure 4.*

In the cross-section at 1550N, the hanging-wall reef extends via link 1a from shear F to shear G and thence downwards to complete the vein loop. The foot-wall member, which is the main reef, passes from shear F to shear G via link 1, extends downwards on shear G and leaves this shear near No.8 level on link 2.

The relationship between the hanging-wall and main veins is evident in the mapping of the upper levels – No.1 level to No.5 level in *Plate 1B.* The intersection of the two reefs, line A-A in Plates 1B and 1C, pitches steeply south from the surface to No.5 level and then swings to a flat pitch along and near No.6 level. *Plate 1C* shows that the higher-grade ore follows this intersection from No.4 level towards the surface. It was this ore-showing that led the Mararoa company downwards to the main orebody.

Farther north, section 1950N, is a foot-wall loop vein where link 1b joins from shear F to shear G. It is seen also on the section at 1550N. A similar foot-wall loop occurs in the Phoenix shaft section, 1250N. The course of the intersection of this foot-wall vein with the main reef is indicated by the dotted line marked B-B in Plates 1B and 1C. It is not always

4. *H. J. C. Conolly, The Phoenix Prospect, Norseman, W.A., Report 55 T/1, April 1934.*

specific, as the foot-wall vein is known imperfectly over parts of the area.

Taken together the courses of these two reef intersections outline the main area of good grade ore on the No.2a orebody, *Plate 1C*. As somewhat similar occurrences are known on the Norseman and Princess Royal reefs, this seemed at the time to be an important feature. In later years, it was found that strong development of loop veins on the Mararoa reef is limited to the No.2a Orebody, as is the extension of ore towards the surface.

Of more general interest is the overall pattern of shear-link occurrence on the main vein. This is shown in *Plate 1B*. The upper limit of link 1 – its intersection with shear F – follows No.4 level throughout most of the length of the No.2a Orebody. It then drops through No.5 level, then No.6 level and finally No.7 intermediate level going northwards. Its lower limit over the area in which it is known, follows a parallel course.

At the southern end of the mine shear F joins with shear G and so link 1 is removed from the vein pattern. Below this position link 2 develops between shear F and shear H, section 1050N. The intersection of link 2 with shear F (and farther north with shear G) is parallel to link 1.

Plate 1C superimposes the shear-link pattern on the ore outlines of Plate 1A. The main ore occurrence follows link 1 and is terminated upwards by its intersection with shear F (except in the area where the hanging-wall vein occurs). Shear F becomes abnormally steep in the northern part of the mine, section 2550N. Ore does extend upwards on shear F at the extreme north end but here the shear is locally flatter (section 2950N).

In his report on the Phoenix Prospect, Conolly described and analysed these features. The following excerpts are taken from his opening summary. "At the Phoenix (i.e., on the Mararoa reef), the gold concentration is controlled by discontinuous horizontal structures, the structures being part and parcel of the reverse fault movement. Flat shoots extend and repeat northwards and southwards. They bottom suddenly down the dip to renew and enlarge again at lower horizons. Disregarding this essential discontinuity, mining (by the old Mararoa company) halted a little beyond the boundaries of the ore first discovered. At the Phoenix the extensions possible have not been tested, nor are they measured by the disappointing results of the mine's lower levels. This optimistic view of possible ore missed includes the Norseman reef. Indeed there are thousands of feet of flat dipping gold bearing reef with favourable structures not explored. The field could include several Phoenix orebodies. It is possible even to visualise a large producer".

A more detailed statement of his views is given below:

"The orebody developed by the Mararoa company is mostly worked out. Ore has been left 'underfoot' at nine places, the important six of which aggregate a length of 800 feet averaging 42 inches wide and 9 dwts. assay ..."

"The gold contours of Plan 2 *(Plate 1A)* show a number of long horizontal shoots ... the top of the orebody parallels the surface some 200ft below the outcrop. Near the shaft the payable quartz bottomed flatly below the 7 level. Below again on the No.8 another shoot comes in ..."

"The distribution of gold is closely related to the vein structure. The concentration of the metal occurs chiefly in the neighbourhood of shear-link junctions.* The spread of

* *This statement applies more to Conolly's interpretation than to the recent re-interpretation. For instance, in his study link 1 was divided into upper and lower links separated by a minor shear. There were other differences in the treatment of the data.*

values is profoundly affected by the elliptical course of vein intersections ... The horizontal discontinuous shear-link structures determine horizontal discontinuous ore shoots, even to the detail of flat north and flat south pitches with change of strike".

"The study thus affords a convincing check of the gold contour observations. It shows the metal to be localised by the structures which must repeat throughout the length and depth of these reverse fault reefs".

Conolly's proposed exploration programme was based on the concepts given below:

1. The exploration must be spread widely over the reefs ...
2. The testing should be systematic with wide spacing from north to south and narrow spacing from east to west.
3. The exploratory drive is unsatisfactory ...
4. The exploratory winze cross-cuts (i.e. crosses) the gold concentration exactly as required ...
5. Systematic drilling could not have missed the known ore-shoots. ... Further, drilling best serves the wide spread of the attack and, if successful, could very speedily decide the option.

His exploration programme is shown in broad outline in *Plate 1C*. It was in fact under way before his report was written.

It used winzing at 900N, 2150N and 2650N. These winzes were supplemented by cross-cutting into the hanging-wall at No.8 level at 1650N and at No.7 intermediate level at 3100N to give bases for testing by underground diamond drilling, the purpose being to get a greater reach down the reef than is possible by winzing. Surface diamond drilling was used at depth and to the north of the old mine workings.

There were good reasons to be optimistic as some earlier diamond drilling had given good results some distance down the vein below the old mine workings – *Plate 1C*. Two holes, M11 and M12, drilled by the old Mararoa company, showed respectively 0.8m at 14.3 g/t and 0.2m at 12.2 g/t. Below the northern end of the mine, a hole (G1) drilled some years later by the Western Australian Mines Department gave 1.9m at 11.8 g/t.

Vein Contouring

Towards the end of 1934, development results were coming in. There were some good results but some poor ones also – see *Figure 5A*.

The winze at 900N had shown quartz but no gold. The winzes at 2250N and 2650N had shown good reef but the assay values were erratic and relatively low grade.

Two of the deep surface diamond drill holes showed good quartz widths but the gold assays were low. Hole No.8 at 750N had given 1.5m assaying trace and Hole No.6 at 2150N had shown 2.4m at 0.9 g/t. A bright spot was Hole No.7 at 3600N which had intersected 1.4m at 15.5 g/t and had suggested therefore a continuation northwards of the orebody. Hole No.9, 150 ft to the east, had shown only two narrow quartz veins, both assaying trace, but this was not necessarily discouraging. Poor results are not uncommon at Norseman, even within areas of good ore occurrence.

The results of the two underground diamond drilling traverses are given in *Figure 5B*.

The drilling at 1650N was wholly discouraging and suggested, in conjunction with surface hole No.8, that the good results in Mararoa holes No.11 and No.12 had not indicated a significant position at depth.

The 3100N traverse had fared better. The underground diamond drilling had given acceptable results and there was great encouragement in surface diamond drill hole No.11 which had shown 2.5m at 14.0 g/t. At the time, this was the deepest ore known in the field.

The time for decision on the Phoenix option was not far away and Conolly faced the problem of advising on the ore reserves indicated by the exploration programme. More realistically, he had to make his best guess at the ore potential.

He was not satisfied that his shear-link study had given sufficient guidance and this led him to the idea of contouring the Mararoa reef in relation to an inclined plane approximating to its average strike and dip.(5) This technique presents changes in vein attitude and position as a series of ridges and valleys, domes and basins in a manner similar to that given by the contouring of land surfaces. More importantly, it is three-dimensional and so gives better methods of comparison and analysis than are possible with conventional plans and cross-sections.

The basis of the method and its results are given in *Figure 6*. The inclined plane had a strike of 1° 50' west of north and a dip of 42° 30' east. Measuring vertically from the inclined plane to the reef was for convenience *(Figure 6A)* and for ease of plotting in plan projection *(Figure 6B)*. These measurements were made throughout the area of the old Mararoa company workings, giving a mass of detailed data. This was supplemented by a scattering of points close at hand from the exploratory winzes and farther afield from the diamond drilling results.

The contouring of the data is given in *Figure 6C*. The contours are shown as full lines in areas of good data and as dashed lines where the information is meagre and scattered. Dashed lines are predominant, particularly northwards and in depth. The contouring method was used to project the known into poorly-known territory.

Figure 6C shows a set of ridges and valleys trending north-south, or nearly horizontally, crossed by another series trending east-west. For convenience in reference, the horizontal ridges are designated R1, R2 and the valleys V1, V2. The main east-west features are described by their co-ordinate positions.

In the central part of *Figure 6C*, V1 and R1, the 1650N ridge and the 2400N valley are dominant features producing a strong dome at 1650N and an equally strong basin at 2400N. The strengths of these features are evident in the cross-sections at 1650N and 2400N – *Figure 6D*. The 3100N ridge is strong at the surface but weakens at depth. At this position, V1 is barely discernible and R1 is a very slight feature as can be seen from the contours and the cross-section at 3100N. The other horizontal features, V2 and R2, are based more on interpretation than on detailed knowledge.

In Plate 2A, these contours are compared with the known orebodies of the old Mararoa workings. A key feature is the even distribution of the No.2a Orebody across the 1650N ridge although with more weight to the south than to the north. The extension of ore towards the surface lies wholly to the south of this ridge. In the other direction, most of the ore (and the best of it) occurs between V1 and R1, i.e., on the flatter dipping

5. *H. J. C. Conolly, A Contour Method of Revealing Vein Structures, Economic Geology, Vol. 31, pp 259-271, 1936.*

section of the reef – link 1 in *Plate 1C*. Ore continues below this position, although somewhat raggedly, between R1 and R2.

To the north, the No.2b Orebody gives a limited base for comparison. Quite clearly, V1 relates to the top of the orebody. The winzes at 2250N and 2650N give some indication that ore extends downwards towards R1. In the north-south direction, this ore lies mostly to the south of the 3100N ridge but is balanced perhaps by the good ore to the north of it in surface hole No. 7. *(Figure 5A)*.

The strong valley at 2400N causes some interruption in the horizontal spread of ore and particularly in its quality (compare *Plate 2A* with outlines of ore better than 12 g/t in *Plate 1A*).

Conolly's general conclusion was that ore favours the 1650N and 3100N ridges, spreading up and down dip along them and in their vicinity. If anything, there is a bias towards their southern sides. Away from them, i.e. approaching the east-west valleys, the shear-link structure of the vein becomes dominant and the ore follows the horizontal features V1, R1, etc.

Using this thinking as his guide, he outlined his expectation of new ore, chiefly around the 3100N ridge. This is shown in *Plate 2B*. There was a minor addition (theoretical) below No. 8 level at the 1650N ridge and a positive addition farther south, where drilling and subsequent development had shown that the old company had driven a minor hanging-wall vein on No. 8 level and so had missed the main channel in its foot-wall.

Conolly was convinced that his interpretation was right. In his report K/53[6] he stated "The ore found totals 230,000 tons with a recovery value of 7 dwts per ton. I advise and am confident that this prospect warrants the formation of a company to explore the district and to start mining of the Mararoa reef ... the ore zone will probably extend, pitching flatly north, at least to 5000N" (i.e. to 1400 ft beyond surface hole No. 7).

His confidence was soon to receive a shock – a severe one. Early in 1935, a visit was made to Norseman by Dr. Donald McLaughlin on behalf of the American partners in Western Mining Corporation, Case Pomeroy. McLaughlin had doubts about Conolly's evaluation and requested additional diamond drilling to test the postulated ore position astride the 3100N ridge.

The results were disastrous – see *Plate 2C*. The key failure was hole 74 which showed no ore, and in fact no quartz, between the 3100N ridge and the good result in surface diamond drill hole No. 7.

However better tidings were not far away. Soon after McLaughlin's visit, it was realised that diamond drill hole No. 59 from the 1650N cross-cut may have intersected a minor hanging-wall vein and that the main reef could be at a deeper position. To test this possibility, hole 59 was deepened and in May 1935 gave the excellent result of 1.3m at 16.7 g/t – *Figure 7*.

This led Conolly to a new hypothesis. The hole 59 result crystallised an idea that had been around for a year or so – that ore occurs in depth as a succession of zones. The underlying thought was based on the observed tendency of links to occur one below the other in the cross-sections of the mine. It would be possible that this pattern could be repeated on a larger scale, i.e., that an ore-bearing zone strong in link structures could be

6. *H. J. C. Conolly, The Norseman Prospect, Report K/53, 18/1/1935*

followed in depth by a steeper zone where shearing is predominant, to be succeeded in turn by another link-rich section and so on.

The deeper ore-zone concept is shown in *Plate 2C,* together with a change in the pattern of horizontally-disposed ridges and valleys. This change presented no problems. Ridge R1 and valley V2 were terminated at 1650N and ridge R2 was joined to what had been ridge R1 at 3100N. Redrawing the contours was a simple matter as the data in this area were sparse indeed.

The result in hole 59 and the deeper ore-zone concept gave new heart to the project. However a long job of shaft-sinking followed by driving on the No.16 level was needed to reach the hole No. 59 position. In consequence, it offered little chance of an early supply of much-needed ore.

The northern end of the mine seemed likely to be more helpful. During 1935, a winze was started at 3050N as a follow-up to the drilling results obtained at 3100N. It was in fact an internal shaft and was equipped as such, being used for many years to commence the deeper levels at the northern end of the mine.

The first new level to be driven was No.8 – northwards from the old mine workings and southwards and northwards from the 3050N winze. It was expected to yield a good supply of ore extending up to the old No.7 intermediate level. However, the results were disappointing. The drive and subsequent stoping above it showed ore occurrence to be patchy and generally poor in grade.

The position reached by March 1938 is shown in *Figure 8A* in relation to stoping outlines as they are known today. The hanging and foot walls of the main shaft had been closely diamond drilled but had shown no ore despite the fact that the shaft is near a significant orebody lying to its south. The 3050N winze was really ill-starred. It had gone well past the good result obtained by surface drill hole No. 11 and had shown nothing of consequence. Subsequent work showed good ore close by to the north and the south but this was no consolation at the time. The only bright spot was the beginnings of some good results on the No. 9 level in the vicinity of surface drill hole No. 7.

By the end of 1937, it was apparent that a re-appraisal was needed and the preliminary work, prior to an evaluation by Conolly, was done by the writer in the early months of 1938. The new workings were mapped, the 1934 shear-link study was extended and the vein contouring versus an inclined plane was up-dated. There was an increase in knowledge but no new light – basically because the area of new information was not sufficient to be helpful. Terence Conolly had once again to make a major recommendation on inadequate data.

Despite the disappointments, his faith in the Norseman project had not wavered. There was no report as such on the 1938 studies. Instead, Conolly wrote a succession of memoranda, thirteen in all; some were extensive, others brief. The burden of them was that the prospects were still good and that there was a need to increase effort and to concentrate on the areas that mattered most. The key memorandum, No. 12, dated May 1938, set out a detailed programme of exploration and development.

Fortunately the wheel of fortune took a kindly turn soon afterwards and better results began to flow during the latter half of 1938. Conolly had been right in being optimistic.

The Ajax Shaft

The position at December 1938 is shown in *Figure 8B*, again in relation to stoping outlines as they are known today.

Driving from the 3050N winze was at last showing some encouragement and it had become evident that an ore position of some consequence was possible. In the main shaft area, the No. 12 level south was giving indications of an ore position which proved ultimately to be useful but not of very good grade. The important point was that new ore was being developed at depth, despite the poor result in surface drill hole No. 8. The No. 16 level had reached the position of drill hole 59 but grade and continuity were less than expected. However this did not matter – again the significant factor was that ore was being indicated at depth.

These results seemed to be a full confirmation of the deeper ore-zone hypothesis. The concepts, as they stood then, are shown in *Figure 9*. The ore zone containing the old Mararoa company orebody and its newly found lateral extensions lies on a flatly-dipping section which terminates upwards on steep and tight shearing and downwards on another steep shear, *Figure 9A*. Below this shear, the lower ore-zone occurs in another flatly-dipping section of the reef to be terminated, in all probability, by another steep shear at greater depth.

The underlying idea had always been that sketched in cross-section in *Figure 9B*. The presumption is that the originating compressive stress was still active, or slightly so, at the time of quartz deposition. In consequence, the steep shears were under tight pressure and the flatter dipping sections containing the links were relatively open, as is indicated in the cross-section. The flatter dipping sections were more favourable therefore for infilling by quartz and the deposition of gold.

A new element was recognised at this time – the indications of new ore to the south of the main shaft suggested an overall south pitch, as is shown by an arrow in *Figure 9A*. A south pitch for individual ore shoots had always been evident on the Norseman reef and also at Princess Royal. Further, it had been established by then that the pitch of the Butterfly orebody to the south of the old Mararoa mine is clearly to the south.

A parallel observation was that ore favours the flatter-dipping sections of the reef having an east-of-north strike. The effect on pitch is sketched in *Figure 9C* which shows a composite plan view of a hypothetical change in reef dip and strike at two positions say 100 ft apart vertically. The intersection of the two reef planes will pitch south as is indicated by the arrow. This new thinking led to the east-west ridges and valleys of the vein contouring studies being changed to controlling features pitching southerly.

The south-pitching element crosses from the upper to the lower ore zone – the arrow in *Figure 9A*. This new factor and the improved development results seemed to have given a solid basis to the geological concepts. However, the financial position was poor as the mine had a large and increasing debt to the Gold Exploration and Finance Company of Australia, based in London. As this company owned most of the shares in Western Mining Corporation and provided its finance, it controlled the general direction of the operations of W.M.C. and hence those of Central Norseman. To improve the financial position, the G.E.F.C.A. board wished to push ahead as fast as possible at Norseman and proposed a new deep shaft as the best way of doing it. Inevitably this new shaft, called the

Ajax, had to be vertical as the technical advice to G.E.F.C.A. came in the main from British-South African engineers whose experience had been mostly in and around Johannesburg.

There is no record of the thinking behind the Ajax shaft project. However, it is more likely than not that Terence Conolly had a major influence as he was in England towards the end of 1938 following a visit to the U.S.A. and Canada. The technical advisors to G.E.F.C.A. would have consulted him for guidance in their project and on the best placement of their proposed shaft.

It is easy to envisage the basis of the project. The concept would have been a succession of ore-zones, one below the other, coupled with the newly-recognised south-pitching influence. The presumed basis is given in *Figure 10, 10A* for the Mararoa reef and *10B* for the Norseman reef. The Ajax shaft covered the known ore on the Mararoa reef in the Central Norseman ground plus any that might be discovered farther north; also, it was well placed to command the Mararoa and Norseman reefs to the east of the leases of Norseman Gold Mines.

There was ample room in depth for new discoveries. As is indicated in the sketch cross-section given in *Figure 2*, the Mararoa reef would intersect the Noganyer sediments at a vertical depth of about 3000 ft, or slightly more than 4000 ft down the dip of the vein. The two known positions occupy 2000 ft or so down the dip; the remaining 2000 ft offered ample scope for a third ore-zone and a good chance of sufficient space for a fourth repetition. There was room also on the Norseman reef for significant ore at depth.

It was an attractive idea with great possibilities but unfortunately it was wrong. The first sign of trouble came at the beginning of 1940 when the Ajax shaft passed through the Norseman reef – *Figure 11A*. It proved to be weak and dispersed, showing two channels of no real consequence. Soon afterwards the shaft passed through a slate zone, previously unknown.

The shaft was continued and a cross-cut at the No. 22 level was driven west towards the Mararoa reef. En route, it passed through the slate zone. The Mararoa reef proved to be as weak and dispersed as is the Norseman reef. Farther north, the Phoenix shaft had been extended to No. 26 level *(Figure 11B)* but showed that the Mararoa reef had broken up below the No. 22 level and that its location at the No. 26 level was uncertain, if it existed at all in any strength.

Prospecting at depth had failed and it was clear that there was something wrong with the geological concepts.

The Favourable Bed Hypothesis

It fell to the lot of Haddon King to find an answer. At the beginning of 1936, he had been appointed Chief Geologist of Western Mining Corporation replacing Terence Conolly who had moved on to become geological adviser to the W.S. Robinson group of companies. Conolly's new position included acting as a consultant to Western Mining Corporation. Haddon had relatively little contact with Central Norseman until the middle of 1938 as Norseman had remained a special interest for Conolly and he had continued to advise on activities there. King's involvement became considerable and then total when the Ajax shaft project got underway.

The appearance of west-dipping slates in the Ajax shaft prompted a search of the surface and this soon led to the Agnes Venture reef. This reef had been found in the early days of the field and consists of quartz veins of little value following the west-dipping bedding of a slate band. It was clear that the slate occurrence in the Ajax shaft relates to this Agnes Venture slate.

The slate-bearing territory was followed northwards and southwards on the surface and was found to contain semi-continuous but scattered outcrops of slate dipping to the west and trending a little to the east of north. This led to the view that it would be advisable to map the sediment belt to the east to get a better view of the overall geological setting of the Norseman field. These sediments and the field as a whole had been mapped in earlier years by W.D. Campbell of the Geological Survey of Western Australia[(1)] but this mapping did not give enough detail for King's needs.

The new mapping of the field was done by H.W.B. Talbot*. Its value proved to lie in the tracing of the Agnes Venture slates and in the detailed mapping of the eastern-most sequence, the Noganyer Group. This group contains numerous and extensive bands of ferruginous fine-grained sediments which at the surface occur as banded iron formation or B.I.F., a hard material yielding strong outcrops. B.I.F.'s can be followed easily and, being finely bedded, show abundant occurrences of minor folding.

Talbot's mapping is the basis of Figure 1 and the comments on it. He followed the Noganyer sediments and their B.I.F.'s from the norite dyke on the north to the old Dundas townsite, some 20 kms or so to the south of Norseman. His work provided a good background of general knowledge but did little to resolve the problems encountered in the Ajax shaft. The solution lay in the spatial relationship between the Mararoa reef and the Agnes Venture sequence of slates. This is shown in *Figure 12.*

The Mararoa reef strikes slightly west of north and dips east at 45° or slightly less; the Agnes Venture slate sequence trends a little to the east of north and dips west at about 50°. This divergence in strike and the opposing dips find expression in the depths at which the slates intersect the Mararoa reef. At the Ajax shaft, the intersection with the probable main fracture is at a depth of 1820 ft vertically; 2500 ft farther north, a surface diamond drill hole shows it at a vertical depth of 2180 ft, i.e. 360 ft deeper than in the Ajax shaft. This gives a northerly pitch of 8°.

Relating trends in plan view tells the story more clearly. In *Figure 12* the upper limit of ore occurrence is in general sympathy with the trend of the Agnes Venture line of outcrops and the lower limit of ore is clearly parallel to it. Such observations led Haddon King to conclude that ore on the Mararoa reef occurs in a particular segment of the basaltic sequence, the Mararoa "favourable bed".[(7)] The postulate is that a particular part of the sequence is favourable to strong fracturing and ore occurrence. Above this "bed" (up-dip on the reef) and below it (down-dip) the country rock, although seemingly similar, is less favourable and so fracturing becomes weak and dispersed.

Figure 12 uses present-day knowledge, much of it not available at the time of Haddon King's study. The Agnes Venture slates were known to him at depth only in the Ajax shaft

* *Senior geologist with Western Mining since its inception. He was senior in age also; he was in his sixties whereas the rest of the geological staff were in their twenties and thirties.*

1. *W. D. Campbell, G.S.W.A., Bulletin No. 21, 1906.*

7. *H. F. King, Geological Structures and Ore Occurrence at Norseman, West Australia, Report K/555, 24/4/1941.*

section. The Phoenix shaft was no help as the slates had not been found at that position and their location is still unknown. He had to rely therefore on general trends for both the Agnes Venture slates and the limits of ore occurrence. The upper limit of ore was clear but the lower limit was uncertain. It is a tribute to his analysis that his favourable bed hypothesis has withstood the test of time, virtually unchanged; it is as useful today as it was when formulated.

A curious feature, on other reefs at Norseman as well as the Mararoa, is that the reef fractures occur within wide zones of actinolite schist dipping east at about the same angles as the reefs. There is virtually no contact therefore between the reef fractures and the basaltic country rock; despite this, the overall trend of ore occurrence reflects the layering of the original rock sequence, not only for the Mararoa "favourable bed" but also for a later discovery, the Blue Bird sequence. This seems beyond explanation. However, the favourable bed hypothesis seems valid for all ore occurrences at Norseman. This extends even to the occasional cross reefs (reefs striking east-west and dipping steeply); they carry ore in either the Mararoa bed or the Blue Bird sequence, as the case may be.

Diffusion of Fractures

As is evident in *Figure 3D*, shears of opposing dips cannot pass through one another. One direction is dominant and has both flowage and finite movement; in the other direction, movement occurs as rock flowage only. In the aggregate, this flowage equals the total movement of the opposing flowage and finite shears and can cause a steepening of dip, as in *Figure 3F.*

In the main area at Norseman, the east-dipping fractures hosting the Mararoa and Norseman reefs lay in the dominant shearing direction; if there was any complementary shearing, it would have been absorbed as rock flowage in the west-dipping lavas of the basaltic host-rocks. At depth, however, the position would change as the reef fractures approached the Agnes Venture slate sequence – a series of slate lenses scattered through volcanic ash, much softer than its enclosing basaltic lavas. This change of environment would be exaggerated by shearing of the Agnes Venture slate zone during the original folding of the area, as is sketched in principle in *Figure 13. Figure 13A* shows three horizontal rock layers with the upper and lower members M and K separated by well-bedded sediments S. M represents the Mararoa basalts containing the Mararoa and other reefs at Norseman, S the Agnes Venture slate sequence and K the Kingswood basalts lying between the Agnes Venture slates and the Noganyer Group of *Figures 1 and 2.* The thickness of S is exaggerated to assist presentation.

In *Figure 13B*, this sequence has been folded into a simple anticline by horizontal compression. The two strong members, M and K, have folded to identical shapes and the intervening sediments have adjusted to these shapes by partial squeezing from the fold limbs into the anticlinal crest. M has moved upwards relatively to K. This movement is considerable and would cause strong shearing along the bedding planes of the sediments. As is sketched in *Figure 13C,* shearing would extend in diminishing degree into the neighbouring basalts along inter-flow boundaries and other lines of weakness. A consequence is that an east-dipping fracture would meet an increasingly difficult environment as it approached the sediments. It would not survive in the sediments as the west-dipping shearing within them would be dominant. The likely result is sketched in *Figure 13C*. The east-dipping fracture breaks into a "horse-tail" and degenerates to rock flowage at and near the contact with the sediments.

Fracture Pattern, S Structures

During the years of the Ajax shaft misadventures, development had gone well in the main Mararoa reef workings. Useful ore had been found to the south of the Phoenix shaft between the No. 10 and No. 16 levels and a major position of considerable length had been developed to the north of it from the No. 11 level to the No. 20 level. The stoping operations resulting from these discoveries are shown in *Figure 12*.

There was an increase in the tonnage of ore treated and a significant improvement in its average grade. This enabled Central Norseman to pay its first dividend, thus allowing its major shareholder, Western Mining Corporation, to pay its maiden dividend of sixpence per share in March 1943 – for both companies, ten years after commencing operations.

Geological work fell largely into limbo during the war years as most of the staff had gone into the forces. Reg Clappison did the best he could but it was beyond the capacity of one man to cover the extensive geological needs of an expanding mine. During 1943, he extended the vein contouring versus an inclined plane into the new workings. However, the vein strike and its average dip had changed markedly in the northern and deeper areas from those known by Conolly in earlier years. This caused problems. As a result, Clappison's laborious task achieved little and the method has not been used since.

Due to Conolly's faith in his contouring method, shear-link interpretation had not been used since 1938. After the war, there seemed little point in applying it to an area approaching exhaustion and, in any case, there were more pressing needs to be met.

To cover this gap, an interpretation has been done recently by the writer. Due to inadequate mapping during the war years, particularly on the cross-sections of the mine, it had to be done on a broad basis. This gives a satisfactory result but there is a lack of detail, significant at the time, in the areas covered by the 1934 and 1938 studies. It is easy to remember the general nature of these interpretations but would be difficult to recreate them with any certainty.

The results of the recent interpretation are shown in *Plate 3*. Shears are identified by letters and links by numbers. In *Plate 3A* the near-surface part of the reef is on shear F and is weak; the reef then leaves it via link 1 which terminates downwards on shear G. Link 2 develops below shear G, to be cut off below at various positions by shears H, J and J' (a split off shear J). Shear G joins shear F at the northern and southern ends of the old Mararoa workings and link 1 is removed from the vein pattern. Link 2 dies out northwards where shear J' joins shear G. At the southern end of the mine, link 3 occurs between shears H and J but fades out where these shears join going northwards.

Link 4 is an unsatisfactory piece of interpretation. There seem to be good reasons for joining its upper portion to link 1 and the lower part to link 2 but these joins are denied by some of the data. Link 4 lies in an area where the differences in dip between the shears and the links are slight.

The lower part of the mine is interpreted more simply – link 5 lying between shears J and L. At its southern end link 5 is broken, somewhat doubtfully, into upper and lower members separated by shear K, a split off shear J. It would be possible to make other such splits, mostly minor, within link 5 but this would achieve little and was not attempted.

Figure 14 gives a series of cross-sections covering the length of *Plate 3.* They show the main reef only as the inclusion of the hanging-wall and foot-wall loop veins would be confusing rather than helpful. At the scale of reproduction, the shears and links seem minute. However, at the scale of mine mapping and of the original interpretations, they are significant features.

Plate 3B compares the shear-link pattern with stoping outlines. The overall relationship of ore occurrence to the links is apparent but it is a little erratic in the upper part of the workings. Links are essential to the occurrence of ore but the relationship is not always specific.

Figure 15 repeats the cross-sections of *Figure 14* in a simpler form. There are three major shears, F, J, and L. The lesser shears and their accompanying links are grouped into flat zones lying between the major shears. The upper flat zone contains shears G and H and links 1 to 4, the combinations varying from south to north. Link 4 is a doubtful member at 3050N. The lower flat zone is link 5 except at the southern end where it is 5a-K-5b.

Plate 4A is similar to *Plate 3B* but has lines added to mark the upper and lower limits of ore occurrence. These are of course the upper and lower limits of the "Mararoa Favourable Bed".

In *Plate 4B,* the ore occurrences and their associated links are grouped into two major positions, the No. 2 Orebody and the No. 3 Orebody. These correspond to the upper and lower flat zones of *Figure 15.* Each ore area consists of a central section, which in fact contains most of the best ore, a northern prolongation and a southern prolongation. These outlines have a crude resemblance to a very open figure S and have been known for many years as S structures.

The No. 2 Orebody has a long northerly prolongation lying mostly above the No. 8 level. Its central section includes most of the old Mararoa company main orebody (No. 2a in *Plate 1),* followed by a short southerly tail extending into the Butterfly lease. The No. 3 orebody has long northern and southern prolongations extending from a strong central position. Ore occurrence follows a single link structure – in contrast to the No. 2 position which relates to a succession of links disposed within a general outline.

These S structures and their relationships to the links are seen best in longitudinal projection, *Figure 16.* In the environs of the Phoenix shaft, the upper margin of the No. 2 Orebody drops from the No. 4 level to No. 10 level; the corresponding drop for the lower margin is from slightly below the No. 8 level to a point near No.16 level. For the No. 3 Orebody, there are similar features astride the 3050N winze – the upper limit drops from the No. 12 level to No. 16 and the lower margin from slightly above No. 14 level to No. 20 level. The southerly pitches of these central sections range from 25° to 40° or so and average about 30°. This contrasts with the nearly horizontal pitches of the northern and southern prolongations of the two orebodies.

The knowledge of S structures grew slowly over the years. The influence of an east-of-north strike was spotted when Hugh McKinstry made a preliminary study of the Norseman reef. A strong S pattern containing a steeply south-pitching link was found soon afterwards in a detailed examination of the Viking mine which lies at the southern end of the Norseman reef. However, its significance was not realised at the time.

A striking example of an S structure is given by the old Princess Royal mine, *Figure 17A*. The steeply south-pitching section of the link strikes east-west and dips steeply south. At its steepest, it has a dip of about 85° south. It yielded high grade ore with abundant free gold; its flatly-pitching southern prolongation extends into the heart of the old workings, also high-grade.

S structures are thought to arise from a hanging-wall – south component (east-block-south) in the reverse-fault movement causing the reef fractures. An analysis is given in *Figure 17*, treating the east-block-south component as a separate entity and being concerned mainly with tension fracturing. *Figure 17* considers first a cube of rock with vertical sides and then moves on to a block of rock dipping 45° east.

Figure 17B shows in plan view the cube of rock stressed by an east-block-south shearing couple. This would cause deformation to a rhombus, as in *Figure 17C*, yielding two sets of shears both dipping vertically. The distortion gives, in effect, the compressive stress C-C of *Figure 17C*. Elongation along the other diagonal would give a tension stress T-T and would cause a tension fracture T, as in *Figure 17D*. This would dip vertically.

The remainder of *Figure 17* considers an equi-dimensional block dipping 45° east and deformed by an east-block-south shearing couple. The lower surface abcd is off-set to the east, as in *Figure 17E*, and would develop tension fracture t. The upper surface would be identical to *Figure 17D* and tension fracture T would relate to t of the lower surface, *Figure 17F*. In consequence, the tension fracture T-t would dip south-easterly at about 60°, as in *Figure 17G*.

On its own, an east-block-south movement along the reef channel would yield links that are relatively more open, as sketched in plan view *Figure 17H*. They would thus be more favourable for in-filling by quartz and the deposition of gold. This is only part of the story as the movement causing the reef fractures was mainly up and down the direction of dip and so the main control would be that sketched in cross-section in *Figure 9B*. The east-block-south influence is therefore an added but minor component within the tension fractures. It reports as a tendency for many of the links to have a strike slightly east-of-north over considerable portions of their lengths. In consequence, ore favours the flatter-dipping sections of the reef having this strike. This is similar to Conolly's conclusion in his vein-contouring studies of the Mararoa reef that ore develops best in flatter-dipping areas lying on the southern sides of the east-west ridges. The difference is that, where the strike is significantly east-of-north, the shear-link intersection pitches south as is indicated in *Figure 9C* and not directly down-dip, as pictured originally by Conolly.

The detection of S structures on the Mararoa reef was not easy as the changes in vein attitude are so slight. This is apparent in *Plate 4B* where the only clear-cut evidence is the east-of-north trend of the No. 16 level over the length of the No. 3 Orebody. The 1938 study confirmed a suspicion that had been in Conolly's mind for some time – that S structures are present in the fine detail of the shear-link pattern. This led, later that year, to the recognition of the south-pitching tendency indicated by the arrows in *Figures 9A and 9C*. However, it was many years before deep development gave proof of his hypothesis. It seems reasonable to think, on the data now available, that the southerly-pitching central section of an S structure relates to a local concentration of the east-block-south influence giving, in effect, a "structural knot" lying at the heart of each of the main orebodies.

The Folding Hypothesis

During the war years, mining on the No. 3 Orebody advanced apace and it became evident that a new source of ore was needed. In consequence, the No. 16 and No. 22 levels were driven northwards in the expectation that the run of orebodies would continue in a regular pattern. However, no ore was found and, worse still, the reef channel became weak and dispersed on both levels not far north of the No. 3 Orebody. The position reached by June 1946 is shown in *Plate 5A*.

A concurrent operation was shallow surface drilling to the north of the mine, as is shown in *Plate 5B*. Most holes showed quartz but assays were erratic, both within the individual intersections and from hole to hole. There was a chance that ore had been found but there was no certainty.

It was evident from the poor results from the driving on the No. 16 and No. 22 levels that something had happened to the course of the Mararoa favourable bed. If it had swung upwards, dashed lines A in *Plate 5B*, the near-surface drill holes would lie in the Mararoa bed. In this case, the Northern Star orebody, nearby on the Norseman reef, would be in the bed also. This possibility is shown in Section 1 of *Figure 18*. If it had swung downwards, dash-dot lines B in Plate 5B, the position would be that shown in Section 2 of *Figure 18*. In this case, the Northern Star orebody and the near-surface diamond drilling would be in a higher bed and the Mararoa bed would lie some distance below it.

The downward swing would keep the trend of the Mararoa bed sympathetic with the line of surface outcrops of the Agnes Venture slates but an upswing of the bed would give a significant disagreement. A downward swing seemed the better answer therefore. However, it had the weakness that it would leave the northern ends of the No. 16 and No. 22 levels in the Mararoa bed and so they should have shown strong reef channels.

This problem was met by assuming that folding has played a significant role in locating the orebodies.[8] This concept is shown in Plate 6A. A key factor is the occurrence to the west of the Mararoa reef of the Blue Bird orebody on the Crown reef. On the favourable bed hypothesis, this would require a Blue Bird bed. Going northwards, folding could have swung this bed eastwards to the positions of the Northern Star orebody and the scattered ore indications found in the near-surface drilling on the Mararoa reef. These hypothetical folds in the Blue Bird bed are shown at the surface in *Plate 6A* and a cross-sectional view of the beds and the reefs is sketched in Section 3 of *Figure 18*.

The effect of these folds on the course of intersection of the Mararoa bed with the Mararoa reef is suggested in *Plate 6A*. It shows closely-spaced folds in the area of the

8. *J. D. Campbell, Norseman Development, Report K/631, 4/6/1946.*

No. 1, No. 2 and No. 3 Orebodies. The large fold needed to swing the Blue Bird bed across to the Northern Star mine would lie to the north of the No. 3 Orebody and would carry the top of the Mararoa bed below the No. 22 level.

To test this hypothesis, the reef was tested by diamond drilling from a cross-cut driven east from the No. 22 level at 6025N. A long haul was expected in the search for a new orebody but, for once at Norseman, fortune smiled and the first three drill holes found ore, *Figure 18 Section 4.* The top of ore is only 100 ft or so down-dip from the vague and discouraging channel found on the No. 22 level.

The No. 4 orebody is a long way north of the Phoenix shaft and a new shaft, the Regent, had to be sunk to service it effectively. This was time-consuming. Ore was in short supply as the No. 3 Orebody was approaching exhaustion and so, to meet urgent needs, the 6050N winze was sunk and equipped as an internal shaft. Much of the development of the 25, 27 and 29 levels was done from it and for several years it handled all the stope production from the area. The ore had to be hauled up the winze, trammed nearly a mile along the tortuous No. 22 level and then hoisted up the Phoenix shaft.

No. 4 proved smaller and lower grade than the orebodies of earlier years. Nevertheless, it was a boon to Central Norseman. It was the backbone of production during the post-war years and played an important role in the survival of the company.

The final stoping outlines on the Mararoa reef are shown in *Plate 6B.* The No. 4 orebody occurs in two halves, to the north and south of the Regent shaft respectively. The southern half is flat-pitching and the reef has a strike that is north-south or perhaps slightly west of north. The northern half seems abnormal as the reef-strike is to the west of north but does show an inflection to a less north-westerly strike in the ore-bearing area, as is evident on the No. 22 level. This part of the orebody has a steep southerly pitch but it is unlikely to be the central section of an S structure. A minor point is that the original exploratory drive on No. 22 level would have found ore – if it had been taken far enough and if it had been possible to do this soon enough.

The upper and lower limits of the four orebodies are shown as dashed lines in Plate 6B. The No. 4 orebody does not sit comfortably – in particular, the lower limit swings away from the general trend of the Agnes Venture slates to the north of the Regent Shaft. The folding hypothesis may or may not be correct; if it is valid, the Mararoa favourable bed would follow a course between the No. 3 and No. 4 orebodies such as that indicated by the dotted lines in *Plate 6B.*

A small orebody was found at higher levels of the Regent shaft, as is shown in *Plate 6B,* but it was a disappointing result for the passage of the Mararoa reef through the presumed Blue Bird bed. Equally disappointing was deep exploration northwards. The deep levels were driven considerable distances north of the Regent shaft but the reef had swung strongly westward and showed little ore, *Figure 19.* The run of orebodies on the Mararoa reef had ended.

PRINCESS ROYAL - NORTH ROYAL OREBODY

The Princess Royal reef was found in the early days of the field and the Princess Royal company was floated in Adelaide in 1895. It commenced ore-treatment in 1897 and ceased operations in 1910, having treated 183,300 tonnes for 152,150 ounces of gold – an average grade of 25.8 g/t.

Other companies tested below the mine, to its south and to its north, without success. Fortunately for Central Norseman, the reef outcrops at the southern end of an extensive sheet of flat-lying Miocene sediments which conceal completely the area to the north of the old mine (see *Figure 1)*. The company operating the old North shaft found the reef but, although it tested to a depth of 150m, it missed an orebody lying a little to its north and not far below its deepest opening.

The Princess Royal reef lies in the northern continuation of the basaltic sequence containing the Mararoa and other reefs at Norseman. It trends north-south and dips to the east at varying angles, very flatly in some areas but mostly averaging 30° to 45° east. It occurs within a band of coarse-grained actinolite-tremolite schist, 30m to 50m wide and dipping to the east at much the same angle as the reef. The schist and the reef dip more or less at a right-angle therefore to the basaltic sequence which, as at Norseman, dips to the west.

Central Norseman gained control of the area at an early stage of its operation and an initial study was done by the writer in 1935. [(9)] As the mine was waterfilled almost to the surface, this had to be an office study using the old mine plans, assay data from the annual reports of the old company and some geological mapping of the deeper levels of the mine by the Geological Survey of Western Australia. [(1)]

It achieved little beyond showing that the reef has a shear-link pattern similar to that of the Mararoa reef, that the ore-shoots have a southerly pitch and that the rich portions of the ore-bodies terminate at and near the 380 ft level. Some surface diamond drilling was done over the next two or three years, mostly to the south of the old mine, with results that were uniformly discouraging.

Princess Royal then lay dormant until Haddon King developed his favourable bed hypothesis to explain ore occurrence on the Mararoa reef. It was a reasonable expectation that the Princess Royal ore-body would relate also to a favourable bed. Haddon assumed that this bed would have a trend similar to that postulated for the northern end of the Mararoa reef and used this to guide diamond drill exploration to the north of the old mine. His lay-out, which has not survived, is reconstructed in Figure 20. Three vertical holes, PRS 5, 6 and 7, were drilled during 1941 and were placed toward the lower part of the presumed favourable zone as its upper reaches had given poor results in the old North shaft workings.

PRS 5 intersected quartz assaying 124 g/t over 1.3m – see *Figure 21A*. However, PRS 6 drilled up-dip showed only quartz stringers of low grade and PRS 7 showed a barren shear in the down-dip position – considerable disappointments after a good beginning. Never-the-less, the position was encouraging as a result such as that in PRS 5 leads normally at Norseman to the finding of an ore-body.

There was then a delay of several years due to war-time shortages of labour and materials. When work resumed, it was decided that the best approach would be from the old Princess Royal Central shaft which was de-watered and re-conditioned so that its

9. *J. D. Campbell, Report on Princess Royal Prospect, 1935.*

1. *W. D. Campbell, G.S.W.A., Bulletin No. 21, 1906.*

No 7 and No 9 levels could be driven north. These would straddle the good intersection in PRS 5, *Figure 21A*, but some 500m of driving was needed on each level to reach the drilling traverse. It was hoped that ore would be found en route on one or both levels.

The reef channel proved difficult to follow and the drives were mostly off-reef, as is indicated in *Figure 21B.* This figure gives the position at December 1946 and shows the results of driving in the vicinity of the diamond drilling traverse. No 9 level showed two short runs of good ore and No 7 gave some encouraging values at its northern end. The reef itself was disappointing as it consisted of irregular and over-lapping lenses rather than a single solid quartz reef.

A winze-rise test between the two levels found the reef to be erratic, difficult to follow and mostly poor in grade. A stope above No 9 level sought good ore near the PRS 5 intersection but had only moderate success. Over-all, the indications were that the area contained no more than scattered patches of good grade ore. It was encouraging that these were found on a flatly-dipping section of reef and it was felt that an ore-body should not be too far away – above or below, or to the north or less probably to the south.

As a first step in further testing, the connection between No 7 and No 9 level was extended by rising on the reef above No 7 level and by winzing on it below No 9 level, *Figure 22A.* The results were disappointing as the rise gave a poor result and the winze showed shear and low-grade quartz in a reef that turned to a steep dip below No 9 level. However, another rise above No 7 level a little farther to the south gave an encouraging result averaging a width of 1.5 m and a grade of 18 g/t over a length of 36m in a reef showing a rather steep dip. It led ultimately to a good stoping position.

The possibilities to the north were tested by continuing No 7 and No 9 levels and by surface diamond drilling. The position at November 1947 is given in *Figure 22A.*

The results of the surface diamond drilling are shown in *Figure 22B.* PRS 8 was encouraging but not convincing as the good assays came from quartz stringers in shear zones yielding poor core recovery. The assays in PRS 9, such as they were, came from quartz stringers also. PRS 10 showed a solid quartz reef and so offered the possibility of a significant position.

The continuation of No 9 level showed erratic reef-channel with some quartz, all of it low-grade. No 7 level showed some additional runs of good-grade reef but the development of quartz remained dis-continuous and patchy. It seemed that, although an area of favourable flat dip had been found, the occurrences of both quartz and ore were too sporadic for favourable country. This led to the view that there could be two favourable beds, both steeply-pitching as is indicated in *Figure 22A,* and that the existing development on No 7 and No 9 levels lay in the unfavourable country between them. This steep-pitch concept had its origin in an improved knowledge of surface geology.

The general geology of the Norseman area is given in *Figure 1.* An area of Noganyer Group sediments occurs on the northern side of the norite dyke a little to the east of the limits of this figure, leading Haddon King in his Report K/555 to the conclusion that there had been a substantial north-block-east movement along the dyke. However, H.W.B. Talbot had subsequently continued and extended his mapping of the Norseman area and had found that the Abbot's Hall beds, to the west of Norseman, continue on line to the north of the dyke. It was evident that movement along the dyke was small and that the off-setting across it of the Noganyer beds was related to folding.

The resulting ideas are given in *Figure 23* which repeats *Figure 1*. The Agnes Venture slate sequence has been added and the continuation northwards of the Mararoa favourable bed has been sketched using the slates as a guide. The Mararoa bed is shown as dotted lines to the north of the dyke, off-setting eastwards by folding in sympathy with the trend of the Noganyer beds to the south of the dyke.

It had been observed by Talbot that folding in the Noganyer Group sediments turns to a steep north pitch approaching the norite dyke, suggesting a steep north pitch to the north of the dyke also. A corollary would be steep north pitches for folding of the Mararoa bed in this area and also for the Princess Royal favourable beds lying a little farther to the west.

The two-bed hypothesis of *Figure 22A* is repeated in *Figure 24*. It shows also a suspected S structure which was derived from two observations:[(10)]

1. An east-of-north strike on No 9 level, indicated by a dashed line in *Figure 24*; it was suggested by lateral diamond drilling from the original approach drive which was off reef in this area;

2. Alternate steeply and flatly dipping reef sections, shown in cross-section in *Figure 25A*.

To test this concept, a cross-cut was put out from No 9 level at 1240mS, as is shown in *Figure 24*, and the position below it in the presumed favourable bed was tested by diamond drilling. The results are shown in *Figure 25B*.

The reef continued in depth but it was steeply-dipping, quartz was sparse and assays were low. It was clear that the concept of an S structure was wrong but the two-bed idea was not dis-proved. It was weakened but could not be ignored as the favourable bed hypothesis relied at that time on finding a recognisable ore-trend. It was decided to continue No 7 level to and beyond PRS 10 and to make it the main exploratory opening until a clearer pattern emerged.

Bonanza!

Driving northwards on No 7 level continued during 1948 but progress was slow. At that time, Central Norseman was short of ore and stoping above No 7 level tended to take precedence over extending the drive northwards. Although relatively small, the tonnage from the Princess Royal stopes was helpful as the main producer, the Mararoa No 4 ore-body, was proving lower grade and smaller than expected.

The results of the driving of No 7 level towards PRS 10 were erratic and rather disappointing; however, stoping above it towards the drill hole gave good results. More encouragement came during 1949 when No 7 level entered good ore to the north of the drill hole, making it evident that a position of some consequence was being developed. Also, the pitch problem seemed to have been resolved. The newly-found ore showed a very flat pitch, at least in its own locale, and it seemed likely that this flat pitch-line could extend back to the old Princess Royal mine workings. Its recognition led to a diamond drilling test to the north of the exploratory openings.

Development and diamond drilling results to December 1949 are shown in *Figure*

10. J. D. Campbell & D. O'Driscoll, Princess Royal Exploration, Report K/716, 19/11/1947.

26A. PRS 16 was unsuccessful, showing reef-channel in the expected position but no quartz. However PRS 15 fared better and gave the encouraging result of 0.5m of quartz assaying 35 g/t. Diamond drilling then moved up-dip from PRS 15, as is shown in cross-section in *Figure 26B*. It gave a satisfactory result in PRS 17 and the spectacular one in PRS 19 of 1.8m at 413 g/t. PRS 20 showed fragments of quartz in very soft oxidised ground and was of course a considerable disappointment. It was found, in later years, to have been misleading.

The approximate position of No 7 level given by the successful diamond drilling traverse is shown in *Figure 26A*. The overall trend of the ore indications is parallel to this level. This gives a very flat pitch indeed, flatter than that used by Haddon King and much flatter of course than was postulated in *Figure 24*.

In the space of a few months, the drilling traverse at 10mN had changed uncertainty into optimism. There was a good chance that ore would be continuous or at least semi-continuous from PRS 8 to the PRS 15 traverse. This is a length of 800m, sufficient to indicate a major ore-body. Beyond this, it was hoped that ore would continue northwards, despite the poor result in PRS 16. It was decided to sink a new underlay shaft, to be called the North Royal, near the successful drilling traverse to develop the area and to bring it into production as soon as possible.

The shaft-sink proved difficult as it had to pass through the Miocene sediments and then through a considerable depth of oxidised ground. This oxidised material was extremely soft and very wet – due to proximity to Lake Cowan and its abundant supply of salt water (up to 30% dissolved salts and, in consequence, very unpleasant). These problems led to the first development opening being cut at No 4 level which was placed at the base of the oxidised zone. It lies a short distance above the reef intersection in PRS 19 – see *Figure 26B*. The shaft-sink showed the oxidised zone to be deeper than had been suggested by the diamond drilling traverse.

No 4 level found very high-grade ore but achieved little as the reef dip proved to be nearly horizontal, despite the result in PRS 20. The level had been sited near the top of a small dome in the reef which dipped below the level – to the north, the south, the east and the west. No 4 level did no more than yield short runs of high-grade ore to the north and south of the shaft.

Due to this failure and to the very soft ground above No 4 level, the objective then became the development of deeper openings as quickly as possible. No 5 level was placed only 23m vertically below No 4 level and No 6 level was cut at a position about half-way between No 5 and No 7 levels. No 7 level had been continued while the North Royal shaft was being sunk and had reached the shaft position before No 4 level was developed.

Driving north and south was started as soon as possible on No 5 level and later on No 6 level; No 7 level was continued northwards as soon as it was connected to the new shaft. Driving northwards was not successful as each of the three levels soon met heavily sheared ground in which the reef broke up and disappeared. It was found ultimately that they had entered a strong shear-zone trending north-north-westerly and dipping steeply east. It is a significant regional feature and has been named the Princess Royal fault. The unsuccessful PRS 16 lies to the north of it.

Driving south-wards on No 5 and No 6 levels was most rewarding, particularly on No

5 level which showed a long run of very high-grade ore. Subsequently, No 8 and No 9 levels were driven southwards with good but relatively modest results. In later years, No 10 level was driven southwards to give access to ore going under-foot on No 9 level.

The final assay results for the whole of the mine are shown in summary in *Plate 7.* Because of the length to be covered, more than one kilometre, this plate has been split into two; *Plate 7A* shows the northern and high-grade half of the workings and *Plate 7B* covers the lesser ore occurrences of the southern half.

The average grades of No 4 level, No 5 level and No 6 level in *Plate 7A* speak for themselves. This high-grade ore was found subsequently to extend most of the way down to No 7 level, despite the relatively poor results of this level. This ore has a length exceeding 400m and a maximum extent east-west of 150m. All of it is good grade, much of it is high grade and a significant part is very high grade, in particular above and below No 5 level.

The effects on ore-grade and on the fortunes of Central Norseman are indicated by the figures given below.

Year	Ore Treated Tonnes	Ore Recovered Grade g/t	Operating Profit
1950	139,900	9.9	$365,000
1951	156,400	8.3	$234,000
1952	151,500	10.1	$418,000
1953	164,000	14.8	$1,234,000
1954	157,800	15.3	$1,030,000

These high grades continued for the next fifteen years, the peak year being 1965 when the average grade was 17.3 g/t and the profit $1,369,000. Present-day equivalents can be derived by recognising that the ten-shilling note of the 1950's had more than ten times the value of today's dollar. Central Norseman had become a very prosperous company – after twenty years of effort, including ten years or so at Princess Royal.

The reef channel was lost about 300 m south of the North Royal shaft on both the No 5 and No 6 levels. They were then driven in waste, as is indicated by the dotted lines in *Plates 7A* and *7B*, towards positions known in the southern end of the mine. The walls were tested by lateral diamond drilling but this was ineffective. It was found later that the reef had moved farther west than could be imagined at the time.

Driving south on No 5 level found almost no ore but No 6 level fared better, finding ore-runs comparable with or slightly better than those seen on No 7 level in earlier years – *Plate 7B.* There are two halves to the ore position – a large and high grade northern orebody and a smaller and lower grade southern section.

The original approach – driving northwards from the Central shaft – was dictated by the complete cover of flat lying Miocene sediments over the area of interest and by the need to develop new ore sources as quickly as possible. It proved to be a poor choice as the narrow span of the southern orebody in the east-west direction made it inevitable that there would be many difficulties.

Shear Link Pattern

As Princess Royal has been so important in the affairs of Central Norseman, it was decided to do a detailed shear-link study of the mine as this allows better and more complete presentation than is given by the mine plans and cross-sections. The vein structures pitch very flatly and so the shear-link pattern is seen best in cross-section; the relatively flat dip allows the over-all view to be given in plan-projection.

Figure 27 gives cross-sections at intervals of 100m from the North Royal shaft to 900mS; these are viewed best from south to north. Sections 900mS to 500mS show normal shear-link patterns, mostly with steep average dips particularly at the southern end. A new feature enters at 600mS where shear M is modified by the introduction of link 9 and shear L. The position is unchanged at 500mS. However, there is a major change at 400mS; link 5 has dropped below No 6 level and link 9 has joined it, moving slightly upwards and enlarging considerably. The junction of the two links has removed shear L from the fracture pattern. Link 5 – link 9, the first flat area of such extent to be found at Norseman, continues in the section at 300mS but disappears to the north of it. It is replaced by a more normal pattern on the three remaining cross-sections. Link 9 continues as a reduced feature related to the re-introduction of shear L whilst, up-dip, new elements appear with the development of shears K and J and links 8 and 7. A minor feature is a foot-wall loop-vein at 100mS.

The one constant component is shear M. It marks the eastern limit of the fracture pattern in all the sections except 900mS, where it has been lost from the system and an enlarged link (link 5 – link 10) joins from shear G to shear N. At the northern end, the Princess Royal fault comes into view without any apparent effect on the fracture pattern.

The large flat area of link 5 – link 9 was not indicated by the initial driving programme and it was some time before it was revealed by stoping above No 7 level. It was missed on No 5 and No 6 levels which had been driven south on a minor hanging-wall vein related to shear K – section 300mS. This failure was no more than an inconvenience at the time as the ore on this flat link lay below No 6 level and had therefore to be mined from No 7 level. The reef was located prior to mining by drilling patterns of the type indicated in the section at 400mS.

The shear-link pattern is shown in plan projection in *Plate 8A*, together with the position of the reef at No 5, No 6, No 7 and No 9 levels (the position of the drive except in areas where it was off-reef). There are minor undulations but over-all the shear-link intersections have nearly horizontal pitches except for link 5 which pitches gently south. North of the shaft, the shear-link pattern terminates against the Princess Royal fault.

The course of No 7 level is of interest. At 450mS, the drive swings sharply to the east (going north); the course of the reef was determined by shear M and the need to provide space for the development of the large flat area on link 9. For similar reasons, the vein at No 6 level swings to the west in this area.

Study of *Plate 8A* shows that except for link 4, all the links are interconnected at one point or another. This is quite different from the pattern on the Mararoa reef where there is a succession of shears and links stepping to the east going north and stepping also to the east going down the dip. This pattern gives the Mararoa orebodies their overall north pitch, a feature seemingly lacking at Princess Royal.

In *Plate 8B*, stoping outlines are super-imposed on the shear-link pattern of *Plate 8A*. In general, ore occurrence favours the links but this relationship is not always specific. From 400mS to 800mS, the southern orebody lies on and near link 5 but farther south follows shear G, with a slight extension upwards onto link 4. It joins to the northern orebody through shears M and L and link 9. The larger and higher grade northern orebody shows its best development where link 9 leads into links 7 and 8. The very high-grade ore relates to link 7 and its attendant shears and, to a lesser extent, to link 8. Stoping extends to the south of the interpreted limit of link 7 on the minor hanging-wall vein related to shear K (see cross-section at 300mS, *Figure 27)*.

There is a small area of stoping above No 9 level at and south of 900mS; its southern tip is the site of the discovery hole PRS 5 (see also *Figure 26A)*. This position is near the centre of the ore-making link 5 but, according to the stoping pattern, the favourable stratigraphy lies higher – at and above No 7 level. The high-grade result in PRS 5 was fortuitous indeed.

Near-Surface Ore

Towards the end of the 1960's, Central Norseman's prosperity faded and in 1971 the company made its first loss since 1939. The causes were rising costs and a low price for gold ($31.45 per ounce in 1970), coupled with a declining grade. A major factor was the exhaustion of high-grade ore at Princess Royal. It was decided in 1971 to cease operations at Norseman after all the available ore reserves had been mined out. The North Royal shaft was closed during 1972.

Fortunately the price of gold commenced to rise during 1973 and it was decided to continue at Norseman. However, a difficult time lay ahead as development had been suspended in 1971 and so new positions had to be explored and developed to re-establish production at an adequate level and an acceptable grade.

The problems are indicated by the figures given below.

Year	Ore Treated Tonnes	Ore Recovered Grade g/t	Operating Profit/(Loss)
1970	163,000	11.3	68,000
1971	153,000	10.7	(148,000)
1972	150,000	10.3	216,000
1973	146,000	8.8	273,000
1974	130,000	7.4	364,000
1975	116,000	7.6	(214,000)
1976	150,000	15.9	1,522,000

The dramatic change for the year ending June 1976 was due to renewed production of high grade ore at Princess Royal.

Production had continued at Princess Royal but there was little ore available and the grade was low. As underground opportunities seemed limited, it was decided to search for new ore by surface drilling between the top of the underground workings and the outcrop of the reef (beneath the Miocene sediments and so unknown). [11] This test started at the southern end of the area of interest with widely spaced traverses and worked northwards. The southern traverses had limited success but results improved to the north. Over several years, the area of interest, extending from 700m south of the North Royal shaft to 500m north of it, was covered by a grid of holes at intervals of 25m north-south and 30m east-west.

The area drilled lay mostly in very soft oxidised ground and had in fact been covered by surface drilling of lesser intensity during the early 1950's. At that time, the main method available for testing soft ground was churn drilling and it proved incapable of giving worth-while indications of any quartz reefs intersected. Fortunately, by the 1970's, drilling techniques had been developed which allow core recovery in all but the softest of ground and so most of the new exploration at Princess Royal was done by diamond drilling. Core recovery was not always good but was sufficient to show that a considerable area of good-grade ore had been found.

Two of the drilling traverses are shown in cross-section in *Figure 28*. The unexpected features are the very flat dip of the main (foot-wall) reef, its great extension westwards and the occurrence of strong hanging-wall reefs relating to the upward extension of shear L.

The cross-section at 25mN contains holes 21, 22, 23 and 24, part of the 1949-50 diamond drilling test – with poor core recovery. These holes show multiple quartz intersections, some of good grade. However, development and mining of them was not possible at that time due to the very soft nature of the oxidised wall-rocks.

These problems remained of course in the 1970's but there was a difference – the depth of oxidation is less away from the environs of the North Royal shaft. The data given by the pattern of surface diamond drill holes allowed underground openings to be guided to areas where the foot-wall reef occurs below the oxidised zone. An area was available at and above No 5 level to the north of the shaft adjacent to the Princess Royal fault; it yielded a considerable tonnage of high grade ore at a crucial time. Another position was on No 4 level, some distance to the south of the North Royal shaft. Operations at and above No 4 level in this area were not rewarding but they did disclose a key element of the Princess Royal fracture pattern – a complex and steeply pitching vein-loop system.

The S Structure

Figure 29A is a plan showing the reef pattern at and around the vein-loop at No 4 level; *Figure 29B* shows No 5 level and *Figure 29C* shows No 6 level. They are interpretations of the data given by the No 4 level drive, the surface diamond drilling programme, underground diamond drilling and, where applicable, the No 5 and No 6 level drives. On No 4 and No 5 levels, shears G & J are joined by link 5 which dips to the south-east at about 50°. In consequence, its intersections with the two shears pitch steeply to the south-east.

As is not uncommon in complex areas, a foot-wall vein developed also. It branches from link 5 along shear H and joins back to shear J along link 6. As this link dips steeply to the south-east, its intersections with the two shears pitch steeply to the south-east, as

11. J. F. Fradd, Chief Geologist, Central Norseman, 1971 – 1976, Personal Communication.

does the intersection of shear H with link 5. No 6 level is different from No 4 and No 5 levels as the hanging-wall reef has faded and only the foot-wall reef remains, with the added feature of link 7 joining from shear J to shear K.

These plans show also the mis-adventures of the No 5 and No 6 levels and their failure to find the reef in this complex area.

These three interpretations are brought together as a composite plan in *Figure 29D* to show their positions relatively to one another and to show also the intersection of the two reefs. This intersection forms the upper part of an incomplete ellipse pitching steeply to the south-east; the convolutions of its northern side relate to the introduction into the system of link 7 between No 5 and No 6 levels.

The central core of the structure is covered by a series of cross-sections in *Figure 30* giving views of the vein-loop in its various forms. Its complexity is evident in the cross sections at 250mS and 275mS. The hanging-wall reef weakens at 300mS and peters out to the south of it, making it evident that the foot-wall vein is the main reef, despite appearances at No 4 and No 5 levels in *Figures 29 A and B.*

Plate 9A shows in plan projection the shear-link pattern of the foot-wall reef in the vein-loop area. *Plate 9B* shows it in longitudinal projection, where a noticeable feature is the turning to a steep pitch of links 6 and 7 in the area of the intersection with the hanging-wall reef. The various elements relating to this intersection are shown in both plates and give the sequence of this changing pattern with more continuity than is possible in the cross sections of *Figure 30.* The vein-loop forms the central core of a strong S structure, as is indicated in both plates. However, it is not a simple structure. On the southern side, shear G peters out and so the upward sweep of its intersection with link 5 peters out also. A consequence is that the upper part of the structure has no defined limit. Shear L, the lower limit of link 5, wedges out going northwards and its place is taken by the large and flat link 9. This link is in fact a key component of the vein-loop structure, not as a member but as a feature affecting its downward sweep.

To the north of the vein-loop, the cross section at 200mS *(Figure 30)* shows a normal shear-link pattern. In the more southerly cross sections, many members of the shear-link pattern of the S structure terminate against link 9, either directly or indirectly through link 6. The result is that links 5, 6, 7, 8 and 9 and their related shears are reduced, on the southern arm of the S structure, to shear G, link 5, shear L and a decreasing link 9. This change in the fracture pattern reports in the occurrence of ore – it is abundant and high in grade to the north of the S structure and somewhat meagre and lower in grade to its south. The best of the near-surface ore lies on the northern arm of the S structure on link 6 – shear H – link 5. Within the vein-loop area, link 6 carries a good width of high-grade ore but Shear H and link 5 performed poorly – an abundance of quartz low in grade.

Figure 31 shows the stoping of the original underground operation and, in a different stipple, the areas stoped in the second period of underground mining. In order of importance, the new areas are:-

1. Above No.5 level to the north of the shaft, adjacent to the Princess Royal fault on a flat link later to be called link 6;
2. Extension upwards through and above No.6 level on link 9 and link 6 in the vein-loop area;

3. Scattered positions at and above No 4 level some distance to the south of the North Royal shaft and at one position to the north of it.

The Open-Cuts

It was evident from the surface diamond drilling data that the depth of the oxidised zone placed severe limits on the use of underground mining and that another method was needed to mine most of the ore indicated by the drilling programme. It was decided therefore to start a trial open-cut to the south of the North Royal shaft. This No 1 Open-cut succeeded beyond expectations as the quantity of ore and its average grade were greater than estimated, in fact considerably greater. It was followed by the No 2 Open-cut at the northern end of ore occurrence. This was successful also.

By the end of the 1970's the supply of under-ground ore was approaching exhaustion and so the North Royal shaft was no longer critical to the operation. It was decided therefore to take a drastic step and to sacrifice its upper reaches to mine the near-surface ore around it. The resulting No 3 Open-cut was the largest, the deepest and the most successful. It was followed by the No 4 Open cut at the southern end of near-surface ore. Currently, the old Princess Royal workings are being mined as the No 5 Open-cut.

The reef areas mined in the open-cuts are shown in *Figure 32, Figure 32A* for the foot-wall reef and *Figure 32B* for the hanging-wall reefs. The latter is a composite of the mining of several hanging-wall reefs, mainly the two reefs shown in *Figure 28A*. The profile of the open-cuts is given in longitudinal projection in *Figure 32C*; they are deep openings – the 90m. depth of the No. 3 Open-cut equals the height of a 30-storey building. The length of the open-cuts is 1.1 km and their maximum width is about 350m.

The relationship of the open-cut mining on the footwall reef to the total area of underground stoping is shown in *Figure 33*. The open-cuts were complementary to the underground operation and there was little open-cut mining of previously stoped areas.

The success of the open-cuts stemmed from a number of factors. Firstly, there are more hanging wall reefs than are seen in the underground section of the mine. Secondly, core recovery in the diamond drill testing was often incomplete and, as high-grade quartz is generally more friable than low-grade, the drilling tended to under-state the true grade of the reefs intersected. Thirdly, the weathering process had removed gold from the reefs and re-deposited much of it in the adjoining clays, mostly where these are rich in iron and manganese; as this material is very soft, much of this gold did not report in the diamond drilling results.[12] Finally, there are many minor quartz veins, shown in the drill intersections in *Figure 28* and *Figure 28A;* these were exposed and mined as the open-cuts progressed downwards and made a significant contribution to the total yield.

Six cross-sections through the open-cuts are given in *Figure 34.* Their locations are shown in *Figure 32.* The reef patterns for the sections at 375mN and 475mS are simplified versions of recorded data. The other reef profiles are derived from the interpretation of scattered points of information; the reefs mined would have been more complex in detail.

As is evident from these cross-sections and from *Figures 32 and 33*, the heart of near-surface ore straddles the North Royal shaft. To the north, it wedges out against the Princess Royal fault whilst, to the south, it weakens as it gets more distant from the ore-controlling influences.

12. G.M.F. Hopkins, Chief Geologist, Operations (W.A.), Personal Communication.

General Structure

Figure 35A shows the reef pattern of the near-surface area as a set of "stacked cross-sections" at north-south intervals of 50m, covering 250mS to 450mN. They show the main (foot-wall) reef only. This construction allows the coordination of cross-sectional data from one section to the next and gives also a measure of progressive changes in their profiles. The individual cross-sections have been placed in their correct relative positions in the east-west direction but each section, going northwards, has been placed 30m vertically above its predecessor.

The working set of "stacked cross-sections" was at 25m intervals and covered the full length of the open cut area. The sources of data were:-

1. The shear-link interpretation of the original underground workings;
2. Information from the more recent near-surface stoping shown in *Figure 31;*
3. Recorded cross-sectional patterns in the open-cuts – in No 2 at the northern end and in No 4 at the southern end;
4. In the intervening areas, spot-points given by the quartz intersections in the surface diamond drill holes.

A problem in item 4 was multiple intersections of quartz in the diamond drill holes. These were handled by trial and error, seeking coordinated shapes and changes of shape through the area covered by the cross-sections. These shapes were then converted to a shear-link pattern – as far as was possible. Taking the section at 100mS as an example, the eastern side of the ridge was split into shear H and link 6, leading thence to shear J, link 7, etc. of the underground interpretation. Its western side was treated as an undulating link, more for convenience than from conviction. Segment 3 could well be a west-dipping shear, complementary to the east-dipping shears of the underground section (see *Figure 3B)*. Alternatively, the undulating reef west of shears J, then K, then L going north, could reflect folding in the host rocks wholly or in part. The western limit is shear D which can be traced, somewhat tenuously, from 200mS to 200mN.

The western side, consisting of shear D and link 2, is stable over the length for which they are known. The ridge-valley feature weakens going north and the whole system moves progressively to the right, i.e., the trend is east-of-north. Segment 3 steepens going north and is nearly vertical at 200mN, making it difficult to consider it a link. Steeply-dipping fracturing is evident at 150mN and is seen also at 200mN. The remnant of link 6 continues and pinches out against the Princess Royal fault to the north of 350mN.

The northern-most cross-sections carry a system of convoluted and somewhat chaotic quartz veins. On sections 300mN and 350mN these are joined, rather doubtfully, to link 6 by a barren east-dipping shear. This is a profound change from the more orderly pattern seen farther south and it seems evident that stresses generated by movement along the Princess Royal fault were dominant in this northern area.

Pitch trends are shown in longitudinal projection in *Figure 35B.* The over-all pitch is horizontal or nearly so. Point A follows the course of link 6 and point B traces the introduction, on a gentle north pitch, of the vein-system of the northern-most open-cut.

Pitch-wise this open-cut is at a higher position than those farther south.

Figure 36 is a plan projection adding the interpretation of the near surface area to the shear-link pattern of the initial underground workings (shown in *Plate 8A)*. Links 6, 7 and 8 merge to form a combined link 6 at and north of the North Royal shaft. It is the main quartz carrier of the area. Its flat attitude is indicated by the meanderings of the vein at No 4 level as it progresses westwards to the north of the shaft. Its variable flat dip allowed the reef to peep up at this level as two small "islands" some distance to the east of the main trend. The one at the shaft position was the down-fall of the original No 4 level – the drive had nowhere to go.

The figure shows the crest of the ridge and the valley-line of the features shown in cross-section in Figure 35A. They converge going northwards and are lost to view to the north of 225mN. They trend east of north as does the main shear-link intersection of the near-surface area, the junction of shear D with link 2.

The reef pattern is shown in cross-section in *Figures 37A and 37B*. Changes in attitude are followed best by progressing from south to north. The shear-link pattern is normal from 900mS to 500mS, the average dip flattening slightly going north. The strong development of link 9, in association respectively with links 5 and 6, is evident at 400mS and 300mS. At section 200mS, this flat dip has translated upwards to link 5 via the central section of the S structure of *Plate 9*. It is of interest that segments 2, 3 and 5 at 200mS have an over-all shape similar to that of link 6 – link 9 at 300mS; the only difference lies in the relative strengths of the ridge-valley features. The significant points at and north of 200mS are the return of link 9 to a more normal size, accretions to link 6 by the successive wedgings-out of shears J, K and H, impingement on the Princess Royal fault and the consequent diminution of the shear-link pattern going northwards.

In *Plate 10A*, the total mining outline on the foot-wall reef (open-cuts plus underground) is super-imposed on the shear-link pattern of *Figure 36*. Shear D marks the western limit of ore and the development of ore on link 2 is rather patchy. The main occurrence of ore in the near-surface area lies to the east of the valley line and much of it is on the northern prolongation of link 6. Link 3 carries the orebody of the No 4 Open-cut.

Plate 10B shows in plan projection the full extent of the S structure presented in part in *Plate 9A*. *Plate 10C* is a longitudinal projection showing more clearly the steep southerly pitch of its central core in relation to the flat pitches of the remainder of the shear-link pattern. The sharp change in the attitude of the Princess Royal fault is most apparent and is due to a marked change in the dip of the reef – see cross-sections 100mN and 00 in *Figure 37A*.

The S structure is seen best in *Plate 10B*. Its southern arm extends southwards on link 5. The intersection of link 2 with shear D marks the western limit of its northern arm but the central section remains incomplete as there is a gap between this intersection and the northern limit of the intersection of link 5 with shear G. The lower limit of the northern arm is a little amorphous in places. It has been taken as the change from the flat dip of the open cut area to the steeper dip of the original underground section of the mine – basically the eastern limit of link 6. These and other imperfections in the S structure seem to relate to the strong development of links, mostly of a flat pitch, in the Princess

Royal fracture pattern; these links occupy the major part of *Plate 10B* and are interconnected at one point or another, except for link 4.

The southern orebody of *Plate 7B* follows link 5 except at its southern-most end where it passes upwards onto shear G and link 4. It is relatively low-grade and weak – often a characteristic of the southern arm of an S structure at Norseman. The steeply-pitching central core was not as productive as would normally be expected – the only stoping of consequence lies on link 6 where it extends upwards towards No 4 level from link 9. The northern arm of the S structure, consisting of link 6 and other flat features to its west, carries the near-surface orebodies. Below it is the high-grade ore of *Plate 7A* following links 7, 8 and 9 and extending northwards to the Princess Royal fault. This lower area yielded more than half of the high-grade ore mined from the North Royal orebody.

Plate 11A repeats the total mining outline and shows the great bulge needed to encompass it within a "favourable bed". The effects of this bulging are evident in the pattern of the major shears, as is shown for key positions in *Plate 11B*. Shear M has been swung eastwards on No 7 level. Shear G was swung westwards on No 5 level and died out on entering the bulge, to be replaced at a higher level by shear D farther to the west (R.L. – 30m is the position of a No 2 level, if there had been one).

This bulge adjoins the Princess Royal fault and its occurrence seems due to stresses generated by movement along the fault. The method is likely to have been folding of the favourable bed. A fold pattern conforming to the general outline of the bulge is sketched in *Plate 12A*; for a bed dipping west, it is west-limb-of-anticline pitching north. The movement required is shown by the two dashed arrows in *Plate 12A,* a direction of movement that is known to have occurred along the Princess Royal fault. This fold pattern is notional; the actual form, if this explanation is correct, is likely to have been an intense crumpling of the bed. These crumples would conform to the stress pattern and so, from crumple to crumple, the bed would progress to the east going north, perhaps with some aggregation in places into larger folds.

In *Plate 12B*, the major elements are brought together to show the S structure, the main features of its disturbed shear pattern and their relationship to the postulated fold-bulge in the favourable bed. The correlation is clear; however, the development of this "structural knot" is likely to have been progressive. On this view, the reef fracture would have developed normally in the earlier stages of deformation, with an S structure and a more or less typical shear-link pattern.

Progressively, as fold-deformation developed, this fracture pattern was modified and to some extent over-ridden by features relating to stresses generated by strong and continuing movement along the Princess Royal fault. One result seems to have been the disturbed fracture pattern of the orebodies mined above No 5 level to the north of the shaft and the strange reef pattern seen in the No 2 Open-cut – see *Figure 35*, sections 300mN to 450mN. Another possible consequence is the off-set in the fracture pattern between shear G and shear D.

A line A-A has been marked in *Plates 12A and 12B.* It is parallel to the general trend of the Princess Royal fault and is drawn through the axis of the vein-loop occurring within the central core of the S structure. Much of the ore and most of the high-grade lie to the north of line A-A; there is a bias however, in both ore-grade and continuity, towards the environs of the fault – a reflection no doubt of the structural disturbance caused by it. The direction of movement along the Princess Royal fault is compatible with the east-block-south stress needed to develop the steeply-pitching core of the S structure – see *Figure 17.*

PRINCESS ROYAL - DEEP AND NORTHERN EXPLORATION

Exploration in Depth

After the war, Central Norseman faced an uncertain future as ore reserves were low and no early relief was in sight. Towards the end of 1946, diamond drilling below No 22 level had indicated the possibility of a No 4 orebody on the Mararoa reef but it was no more than a possibility and some three or four years of development would be needed to get effective production from it. The other main exploration effort, unwatering the Princess Royal Central shaft and driving northwards on the No 7 and No 9 levels, had not fared well – see *Figure 21B*. It became advisable therefore to look for additional positions likely to yield a worth-while supply of new ore. The best of these seemed to be exploration below the old Princess Royal mine.

The high grade part of this mine terminated at the 380ft level (80ft vertically below No 3 level) and deeper testing found only a small patch of low-grade ore below No 5 level – see *Figure 38*. This suggested that the Princess Royal reef had passed out of its favourable bed not far below No 3 level, as is indicated in general terms in *Figure 20A*. However, the reef channel continued below there at sufficient strength to be followed successfully by the old Princess Royal Central Company on their No 7 and No 9 levels. These levels showed intermittent quartz but no ore and suggested that, in contrast to the Mararoa reef, the reef channel had survived in depth despite being in unfavourable country. There seemed a possibility that it could regain definition and perhaps show ore if it should enter another favourable bed lying somewhere below No 9 level.

A factor in this thinking was the East reef. It was found by the old Princess Royal Company to the north of their main shaft, at a position about 80m east of the main reef, and was mined by them over a length of about 100m from the surface to No 3 level. It had been found also on No 9 level in a cross-cut driven eastwards by the old Princess Royal Central Company.

The existence of a hanging-wall shear led to the concept expressed in cross-section in *Figure 38B*. The main part of the orebody relates to the vein-loop near and above No 3 level and extends to the surface on a generally flatter-dipping section of the reef; the minor ore occurrence below No 5 level relates also to a flattening of the dip. Below this, the dip is steep to No 9 level. If the East reef should be part of the overall system of fracturing, the reef channel could return to a flat dip at a position somewhere below No 9 level, as is sketched in principle in *Figure 38B*.

If this view was correct, it was likely that such a flattening would have occurred in a deeper favourable bed. The problem then was the best location for the testing of this possibility. It was decided to follow the projection of a known favourable structure, the old Princess Royal position, down its probable southerly pitch searching for its intersection with the postulated favourable bed. [13]. This concept is shown in *Figure 38A*. It was similar to the Ajax shaft project but with a difference – the Agnes Venture slate sequence was known to be a long way to the east of Princess Royal.

A cross-cut was driven eastwards at the southern end of No 9 level and four exploratory holes were drilled, as is shown in *Figure 38C*. The results were disappointing but not wholly discouraging as a quartz-bearing reef channel continued in depth and showed some minor gold values.

During 1947, it had been recognised that the Mararoa favourable bed continued,

13. *J.D. Campbell, Princess Royal Exploration, Memo to F.F. Espie, 7/8/1946.*

virtually undisturbed, to the north of the norite dyke – see *Figure 23.* The thinking at the time was expressed in Report K/716 [10].

"The Princess Royal shear has considerable strength and has been shown to extend over at least 4,000ft horizontal length and for 2,000ft down the dip. Surface mapping and interpretation suggest that the Mararoa bed may not be far below No 9 level and it is a reasonable expectation that where the very strong Princess Royal shear intersects this bed it may offer the chance of a major orebody."

It was suspected at the time that the old Princess Royal orebody lies in the Blue-Bird favourable bed of *Plate 6* but there was no proof.

The reef profile given in *Figure 38C* had shown an encouraging flattening between the two deepest holes, suggesting that the Mararoa bed may not be far below. The obvious move was to extend the cross-cut and to take the drilling to greater depths. However, this presented problems, as a simple extension eastwards of the drilling traverse could take it outside the presumed pitch extension of the structure carrying the old Princess Royal orebody. Beyond this, it had to be recognised that this pitch could be flatter than that shown in *Figure 38A* and so the potentially favourable position could have been replaced partly or wholly by the norite dyke.

It was decided therefore to move the diamond drilling test farther north. The basis is given in *Figure 39.* The diagram relates to *Figure 24* and was an extension of the thinking on which this figure was based. The underlying idea expressed in *Figure 39* was that the favourable vein pattern thought to have been indicated by the new northern workings on No 7 and No 9 levels could extend flatly southwards into a position likely to be occupied by the Mararoa favourable bed. A third project had been added to the exploration programme relating to *Figure 24* and the three objectives were in fact tested concurrently.

The cross-cut of the old Princess Royal Central Company on No 9 level was extended eastwards and the holes shown in *Figure 40* were drilled. The results proved in the end to be disappointing. A strong shear channel encouraged ever-deeper exploration but the shear remained steep and became even steeper in the deeper holes.

Hole PR 88 showed 11.5m core length of quartz but it was buck-quartz and assayed only a trace. Hole PR 127 offered some encouragement with 1m of quartz assaying 3.2 g/t but it was not supported by the later PR 192 which showed two strong zones of shearing carrying only minor gold. The final major test was PR 179 which did not, however, find the long-sought change to a flat dip and in fact showed no reef channel at all. High hopes and an expensive test had a dismal ending.

PR 179 was drilled during 1951 and, except for the drilling of PR 192, the project lay dormant until 1957. It had become evident by then that the North Royal orebodies occur almost certainly in the Blue-Bird bed and that the best quality and greatest quantity of the newly-found ore lie at the northern end [14]. The resulting ideas are shown in *Figure 41A.*

The Princess Royal fault is shown in this figure as the northern limit of favourable structure as it was regarded at the time as a sheared and stretched fold-limb. This makes little difference to the basic concepts. These were:

1. Ore at North Royal favours proximity to a feature trending north-north-westerly;

10. J. D. Campbell & D. O'Driscoll, Princess Royal Exploration, Report K/716, 19/11/1947.

14. J. D. Campbell, Crown Reef and Princess Royal Exploration, Report K/1096, 14/6/1957.

2. To the south of the two North Royal ore-bodies there is a poor zone, beyond which lies the orebody of the old Princess Royal mine;
3. If the spacing between favourable beds known at Norseman should apply at Princess Royal, the Mararoa bed would lie at no great distance below No 9 level;
4. The underground diamond drilling traverses at 1875mS, 1685mS and 1240mS may have given some intersections within the Mararoa bed but they would lie, more probably than not, in the poor zone between the two areas of ore occurrence.

It was decided therefore to extend by surface diamond drilling the old traverse at 1240mS and later to drill farther north at 800mS. These drill holes are shown in cross-section in *Figure 41B.*

Again the out-come was discouraging. The first three holes at 1240mS showed a strong shear containing occasional quartz stringers but the deepest hole, PRS 95, found no reef channel at all. The two holes at 800mS showed a good shear channel with some quartz stringers, a disappointing result in view of their proximity to the Princess Royal fault.

It is not unlikely that the basaltic lavas of the Mararoa bed extend to the Princess Royal area and so the testing of their probable position was worthwhile. However, they do not carry ore on the Princess Royal reef.

Northern Exploration

During 1949, the drilling traverse shown in *Figure 26B* had given the first indications of the high-grade ore of the North Royal orebody. Hole PRS 16 had been drilled concurrently with hole PRS 15 at a position 370m farther north. It showed shear only but was regarded as encouraging as this shear was strong and lay in a position directly in line with the reef intersected by PRS15.[15] Drilling on this northern traverse was discontinued to allow a rapid follow-up of the good result in PRS 15.

Northern drilling was resumed in 1954 and had an early success in PRS 69 which showed 0.5m of quartz assaying 117 g/t – see *Figure 42.* It was decided to drive No 10 level northwards to test this position. However, this drive showed nothing of interest as did two winzes below the level, one at the drill hole position and another a little farther north. The surface diamond drilling programme had been continued but inter-sections of interest were sparse, as is evident in *Figure 42.* The best result was in PRS 73 which cut a narrow reef showing 0.2m of quartz assaying 37 g/t.

The position was reviewed in 1957[14] and this led to the rather optimistic view, shown in *Figure 42,* that PRS 69 had intersected the reef near the upper limit of a south-pitching orebody. However, as driving south on No.11 level from the more southerly of the two winzes from No.10 level showed only low assays in testing around PRS 69, it seemed likely that its good result was a "sporadic high". Interest in the area lapsed until 1960.

Towards the end of the 1950's, the true nature of the Princess Royal fault had been recognised and it had been established definitely that the North Royal orebodies lie in the northern continuation of the Blue-Bird favourable bed of *Plate 6.* In combination, these factors made it evident that the surface diamond drilling to the north of the fault

15. D. O'Driscoll, Geological Notes for the Period 22/6/1949 to 13/9/1949.

14. J. D. Campbell, Crown Reef and Princess Royal Exploration, Report K/1096, 14/6/1957.

had not been taken deeply enough to cover fully the down-faulted continuation of the rock sequence known to be associated with the Blue-Bird bed. In consequence, deeper holes PRS 101, 105 and 112 were drilled and were followed later by PRS 130 and 131. These holes are shown in plan view in *Figure 43* and, except for PRS 101, in cross-sectional view in *Figure 44.*

The best result was in PRS 105 which showed shearing over a core length of nearly 10m. This zone showed several bands of quartz, with poor core recovery in the intervening sheared greenstone. The best assays from the quartz were 0.2m at 76 g/t near the top of the sheared zone and 0.5m at 5 g/t, and 0.7m at 5 g/t, both near its base.

Geologically, these assays had little meaning as they occurred in buck quartz and, in view of the poor results in the other drill holes, were more likely to be "erratic highs" than indications of a significant orebody. Another unfavourable factor was that the reef-dip had remained uniformly steep, as is evident in the cross-sections of *Figure 44.*

This view was not accepted by management and it was decided to continue sinking the North Royal shaft and to drive northwards on the No.22 level. The reef channel proved difficult to follow, as is indicated by the erratic course of the level in *Figure 43,* and showed nothing of consequence. The contours of the reef channel given in this figure, show a trend slightly to the west of north – generally an unfavourable reef attitude at Norseman.

Near the surface, the reef channel shows a local swing to an east-of-north strike and this area yielded a small stoping operation. It is shown at a larger scale in plan view in *Figure 43B*. In the cross sectional view of *Figure 43C*, the reef channel is steeply-dipping. It was a meagre reward for a major exploration effort. Also, it was clear that the Princess Royal fault marks the northern limit of significant ore occurrence.

NORSEMAN REEF

As the name implies, the Norseman reef was the first discovery of the area. It is known over a length of 2.5 km., the northern half having a strong outcrop in hilly country and the southern half being poorly exposed in a lower-lying and soil-covered area – *Figures 45 A, B and C.* Inevitably, the early prospecting was in the northern half and the discoverer of the reef had the misfortune to peg near the northern end where there is a strong showing of quartz but the occurrence of gold is erratic and weak. The Reward Lease proved singularly un-rewarding. The worth-while orebodies lie towards the southern end – *Figure 45A.*

The reef disappears northwards under an alluvial flat and re-appears farther north as the Northern Star reef, *Figures 48 B and C.* Thus the total known extent is more than 4 kms., making the Norseman the longest reef-outcrop of the district.

The Norseman reef is parallel to the Mararoa and lies about 400m. to the east of it, *Figure 1.* Its trend is more sinuous than that of the Mararoa. The reef dips to the east at an average angle of 40° or so at the southern end but becomes flatter towards the northern end and in the Northern Star mine.

Recorded production to 1933 is given below.

Mine	Tonnes	Ounces	g/t
Old Viking, Mildura, Hardy	88,100	45,560	16.1
Viking	56,600	54,320	29.8
Northern Star	14,200	7,740	17.0
Total	158,900	107,620	21.1

The Old Viking, Mildura and Hardy were worked by Norseman Gold Mines and this was the first significant operation in the Norseman district. The Viking mine was discovered and worked some years later by a local syndicate – a rewarding venture. The Northern Star company was less fortunate as, although the grade was good, the tonnage was small.

The Norseman reef attracted interest from the beginning of the Central Norseman operation but it was evident that any testing of it should await a successful outcome of the work then in progress on the Mararoa reef. However, it was decided to commence as soon as possible with the mapping and interpretation of the Norseman reef workings. This became in fact a study of the Viking and Old Viking mines as the Mildura mine lay in the leases of the second Norseman Gold Mines of the 1930's. The Hardy and other positions to its north were in Central Norseman ground. They were mapped but worthwhile interpretation was not possible as the reef areas exposed by the workings were too small. Cross sections through the Viking, Mildura and Hardy shafts are given in *Figure 46.*

The interpretation of the reef structure in the Viking and old Viking mines showed a typical shear-link pattern. As with the early work on the Mararoa reef, the plans and cross sections showing it have disintegrated or disappeared. It was not worth re-doing this interpretation.

Beyond showing the presence of a horizontally-disposed shear-link pattern in both mines, the main feature to emerge was an S structure which is central to the Viking orebody. The flatly-pitching northern arm lies to the north of the shaft between the -100ft. and -200ft. contours in *Figure 45A*. The central section, pitching steeply south, passes through the shaft between the -200ft. and the -300ft. contours; the flatly-pitching southern arm lies astride the -300ft. contour. However, the term S structure had not been devised at this time.

The overall pattern of ore occurrence was of more interest. The Viking orebody is steeply south-pitching and there is a similar but vague suggestion in the Old Viking mine. The orebodies "bottom" at a relatively shallow depth on a line pitching gently south as is indicated in *Figure 45A*. Beyond this, the keen eye of Hugh McKinstry noted that the orebodies occur on the segments of the reef trending east of north and that trends to the west of north are unfavourable for ore occurrence. The over-all flat pitch of the orebodies on the Norseman reef was a confirmation of Terence Conolly's view that the main feature on the Mararoa reef is a very flat pitch and it was found later that most of its ore lies in areas where the strike is slightly east of north.

During 1935, the No. 5 level of the Viking mine was extended south-wards; it lies slightly above the -300 ft. contour of *Figure 45A*. After several hundred feet of driving, the reef channel became dispersed and difficult to follow and the project was abandoned.

During 1937-38, the Viking shaft was sunk from its bottom near No. 8 level to approximately No. 12 level, without the blessing of Terence Conolly and to his great annoyance. He wished all exploration to be concentrated on the Mararoa reef. The Norseman reef channel became weak and dispersed as sinking progressed and this was in fact a warning, not recognised at the time, of the misfortunes that were to occur a few years later in the Ajax shaft.

Following the 1938 re-study of Norseman, some surface diamond drilling was done in the Hardy shaft area. It is shown in the plan view in *Figure 45B* and in cross section in *Figure 46*. Hole S20 gave the encouraging result of 0.2m of quartz assaying 67 g/t but it proved to be an "erratic high". The Hardy shaft was extended as a winze, as is shown in *Figure 45B*, and an opening was driven to and past the drill intersection. It showed nothing of interest and the Norseman reef was abandoned for the time-being.

After the war, interest in the reef was revived by the favourable bed hypothesis developed by Haddon King in 1941 to explain ore occurrence on the Mararoa reef. It was evident that the Norseman reef orebodies of its southern end occur in the upward prolongation of the Mararoa favourable bed and it was likely that, to the north of the Sydney shaft (opposite the Phoenix shaft on the Mararoa reef), the favourable zone should pitch gently north on the Norseman reef, as on the Mararoa.

Possible relationships to the Mararoa favourable bed for the significant Norseman reef orebodies are sketched in cross section in *Figure 47*. In section A – A, through the Viking shaft, there is a strong off-set between the bottom of the ore position on the Norseman reef and the lower limit of ore on the Mararoa reef – the No. 1 orebody in the Butterfly mine of Norseman Gold Mines. Section B – B, through the Mildura and Ajax shafts, is similar. If the cause was folding, as is sketched in the two cross-sections, this fold would align with the one used in *Plate 6A* to explain the off-set between the No. 1 and No. 2 orebodies on the Mararoa reef.

In section C – C, the Hardy mine lies up-dip from the inferred favourable bed of the

Mararoa No. 2 orebody. In section 580mN, the Mararoa No. 2 and No. 3 orebodies come into play. Their favourable bed positions have been projected up-dip parallel to the dip known for the Agnes Venture slate sequence in the Ajax shaft section. There is no case for projecting the fold pattern of *Plate 6A* as these folds derive from an hypothesis based on indirect data.

The possibilities of the Norseman reef were reviewed by Des O'Driscoll* and his analysis and proposed surface drilling programme were set out in Report K/715 [16]

The basis of his drilling programme was:

1. Going northwards from the Hardy shaft, the Mararoa favourable bed would become progressively deeper and the Norseman reef would lie wholly within it at more northerly positions – *Figure 49 B;*
2. Ore should favour sections of the reef trending east of north;
3. These sections should give favourable zones pitching steeply south, as in *Figure 48 A.*

The first traverse of this new programme was placed half-way between the earlier Hardy shaft and 580mN traverses, *Figure 48A*, as it was felt that these had not given an effective test of the north-easterly trending segment of the reef. A second traverse was needed at coordinate 1135 MN to cover a more northerly section of favourable strike.

The results are shown in cross-section in *Figures 49A* and the presumed relationships to the Mararoa favourable bed in *Figure 49B*. Both traverses showed a normal reef dip, with barren quartz at higher positions and a decline to shear only at depth.

The position was reviewed again by Des O'Driscoll towards the end of 1948 [17] and this led to the drilling of another traverse farther north, *Figure 48B*. The underlying concept invoked the folding hypothesis used in the search for the Mararoa No. 4 orebody – *Plate 6A*. This hypothesis relied in part on the encouragement given by the small but significant Northern Star orebody in showing that ore could occur in these northern latitudes. As the folding hypothesis required the Northern Star orebody to lie in the Blue Bird favourable bed and the general concept seemed to be supported by the finding of the Mararoa No. 4 orebody, an obvious project on the Norseman reef was to test it in the Mararoa favourable bed where it would lie in the downward projection of the north-easterly reef trend of the Northern Star position. The main elements are shown in *Figures 48B and 48C.*

The drilling traverse at 1830mN is shown in cross-section in *Figure 49A* and the probable relationship to the Mararoa favourable bed in *Figure 49B*. For the first time in this project, there was an interesting result on the Norseman reef – hole S35 showed 0.3m of quartz assaying 1.9 g/t. As the surface drilling had become relatively deep, it was decided to continue the test by underground diamond drilling. Hole 680 gave the encouraging result of 0.3m of quartz assaying 11 g/t but any enthusiasm was dampened by the subsequent drilling of hole S36. It showed 0.5m of quartz assaying trace. By then, Princess Royal had become the centre of interest and no further testing was done on the Norseman reef.

16. *D. O'Driscoll, Norseman Reef Drilling Proposals, Report K/715, 14/11/1947.*

17. *D. O'Driscoll, Norseman Reef Drilling, Report K/767, 17/11/1948.*

* *D. O'Driscoll, Senior Mine Geologist 1946 – 53.*

CROWN REEF

Early Mining and Testing

The Crown reef lies about 400m to the west of the Mararoa reef *(see Figure 1)* and was mined in the earlier days of the field in the shallow workings of the Golden Crown shaft. The quartz is narrow but high grade – to 1933, the production from these workings was 1380 tonnes at 45 g/t. During his initial examination of the Norseman area, the Crown attracted the attention of Terence Conolly as it seemed to be the next member, going westerly, of an evenly-spaced system of east-dipping quartz reefs, with the added attraction of a high grade. However, he did not rate it highly and his recommendation for it was – "Diamond drilling from the Phoenix mine levels. Not urgent".

In the middle 1930's, the small Blue Bird company took up the ground to the north of the Golden Crown and found a small but high-grade orebody. It operated from 1936 to 1943 and produced 11,100 tonnes for 25,960 ounces of gold, a grade of 73 g/t. This high grade was due to a fortunate circumstance – the quartz reef was seldom more than 0.2 m to 0.3 m in thickness but the wall rock although soft was firm and so could be removed in a separate operation prior to mining the quartz.

To the north of the Blue Bird, the Empress company tried its luck. It and its successor, Dundas Gold Mines, produced 7,200 tonnes at a recovered grade of 17.6 g/t during the years 1941 to 1950. Other companies held ground to the north of the Empress but found nothing of interest.

The only testing available to Central Norseman was diamond drilling within its own leases down-dip below the Blue Bird orebody. This was unsuccessful – it showed no quartz and little sign of a reef channel.

The workings along the Crown reef are shown in plan projection in *Figure 50A*. The reef trends to the west of north on the surface but swings to a northerly strike in the underground workings of the Blue Bird mine. The main ore occurrence does not reach the surface. The reef dips flatly to the east in marked contrast to the steeper dip of the Mararoa reef – *Figure 50B*. Seemingly, the two reefs should intersect at depth.

The interest of Central Norseman in the Crown reef was quickened when Haddon King developed his favourable bed hypothesis to explain ore occurrence on the Mararoa reef. An obvious corollary is that there could be a Blue Bird "favourable bed" to the west of and stratigraphically higher than the Mararoa bed. Its intersection with the Crown reef would pitch gently north. This concept is shown in *Figure 50A* which shows also the programme of surface diamond drilling needed to make an effective test for such small ore-bodies [18] A wait of nearly ten years had been needed to get control of this prospective ground.

Figure 50C shows the results of the initial drilling traverse at 1525mN. Hole C10 gave an excellent result and C 11 was encouraging. However, holes C 8 and C9 were poor and, over-all, the drilling results did not offer any great encouragement. The drilling programme was discontinued for the time being.

Diamond drilling returned to the Crown reef during 1954 but with a changed programme. The first of the new drilling traverses was at the latitude of the Regent shaft – 2105mN in *Figure 51A*. The diamond drilling results are given in Figure 51 B. Holes C 13 and C 14 showed shear only but C15 showed good grades in two parallel quartz reefs and hole C 16 gave a spectacular result. Deeper testing was done from the Regent shaft

18. D. O'Driscoll, Proposed Blue Bird Drilling, Report K/856, 10/8/1950.

workings, C 23 giving a satisfactory result and C 28 showing shear only.

These high grades were accompanied by good quartz widths – a marked change from the narrow quartz of the 1525mN traverse and of the Blue Bird mine. Diamond drilling was extended northwards and southwards, with traverses at about 200 m intervals as is shown in *Figure 51A*. The results of this diamond drilling are shown in *Figure 51C*.

Development of Crown Reef

The diamond drilling had shown a significant ore-body at and north of the Regent shaft and a lesser position to its south. In the cross section at 2105mN *(Figure 51B),* it was evident that projection to the surface would give a position aligned with the Maloney reef – a zone of erratic quartz veins of low grade; this position could be joined back to the Blue Bird mine and to the Crown reef, as is indicated by the dotted line in *Figure51A*. Contouring of the reef intersections showed sympathy with this line and with the outcrop of the Maloney reef – except for the contours below – 600 ft. to the south of the Regent shaft, where a sharp bend was needed to fit the data.

As is evident in *Figure 51B*, the Crown reef could be developed by cross cutting westwards from the Regent shaft, starting with No. 14 and No. 16 levels. No. 16 level was driven first – in a south-westerly direction in the seemingly shortest distance to the reef. However, there was no reef at the expected position and, as is shown in *Figure 51A*, the cross-cut had to be re-directed towards a known position, the reef intersection in hole C 23. The No. 14 level cross-cut was taken in the safer westerly direction. Further problems arose when driving commenced on the two levels, *Figure 52A*. To the north of the Regent shaft, both levels trended north-easterly but, to the south of the shaft, No. 16 level continued in this same direction. As it seemed likely that the reef would continue southwards on this course, the Maloney reef at the surface could not alter direction and join back to the Crown reef, as had been postulated in *Figure 51A*. There had been a mis-interpretation of the data.

This changed situation is shown in cross section in *Figure 52B*. There is a weak upper reef channel, difficult to detect in the diamond drill core, and a lower reef carrying the good intersections made by holes C 15, C 16 and C 23 and in the diamond drill holes to the north of the Regent shaft. The upper reef was intersected in the diamond drilling traverses to the south of the shaft – 1905mN, 1705mN and the earlier 1525mN – and is the reef worked in the Empress, Blue Bird and Golden Crown mines. The lower position is the Maloney reef, established by No. 16 level as persisting on a north-easterly trend. The initial mis-tie is shown as a dotted line in *Figure 52B*.

This revised interpretation caused a problem, as the work had been described for some time in the company's quarterly and annual reports as being on the Crown reef. The solution was simple : the Crown was re-named the Blue Bird reef and the Maloney became the Crown reef – unfair treatment, perhaps, of Mr. Maloney.

As the earlier testing had been too shallow, a deeper diamond drilling programme was needed to test the southerly potential of the newly-named Crown reef. Two diamond drilling traverses were proposed, one at 1340mN and the other at 910mN[14]. They were wide-ranging, aimed at testing the Blue Bird "favourable bed" and positions above it and also at the possibility that ore could occur at depth in the down-ward projection of the

14. *J. D. Campbell, Crown Reef and Princess Royal Exploration, Report K/1096, 14/6/1957.*

Mararoa "favourable bed". These concepts are shown in plan projection in *Figure 52A* and the diamond drilling results are given in cross section in *Figure 53*.

The first diamond drill hole, C 40 at 1340 mN, was a resounding success and there was some support from C 43 to its east. However, there was nothing further of interest, particularly up-dip where the channel was found to be weak and contained little quartz. The results at 610 mN were uniformly poor and subsequent development showed that holes C 40 and C 43 lie near the southern end of ore occurrence on the Crown reef. The deeper parts of the drilling traverses suggested that significant ore occurrence was unlikely in the Mararoa bed position.

These drilling traverses indicated the general course of the Crown reef from No. 16 to No. 27 levels, as is shown in *Figure 52A*, and it was decided to develop the southern part of the mine using these trends as a guide. Development commenced on levels 14 and 16, followed by 19 and 22 and, when justified, by No. 25 and No. 27 levels. In later years, some deeper levels were developed by cross cutting westwards from the Mararoa reef workings – the positions of interest are well to the south of the Regent shaft.

Reef Patterns and Ore Shoots

The levels driven on the Crown reef are shown in plan projection in *Plate 13A*. Mostly the reef trends N.E. – S.W. – in marked contrast to the northerly to north-westerly course of the No. 22 level drive on the Mararoa reef. To the north of the Regent shaft, however, the Crown reef turns to a northerly to north-westerly strike and intersects the Mararoa reef at no great distance farther north. This intersection is hypothetical rather than actual as both reef channels are zones of weak and dispersed shearing in this area. For this reason, the northern end of the Crown reef drive on No. 22 level is shown as a dashed line instead of a solid one.

Over most of the mine area, the dip of the Crown reef is flatly to the east, as is evident in the series of cross sections in *Figure 54*. These are repeated at half-scale in *Figure 55* to show relationships to the Mararoa reef. A noticeable feature is that the Crown reef steepens as it approaches the Mararoa and becomes parallel to it where they are in close proximity – sections 2301mN and 2500mN.

In the productive area, the Crown reef channel consists in many places of two or more closely-spaced shears with quartz veins occurring on one or more of them, not infrequently as over-lapping lenses which wax and wane up and down dip and along strike. The occurrence of gold shows similar but more variable trans-positions. No. 14 and No. 16 levels show a distinct hanging-wall reef but this is not the norm. Mostly parallel reefs, where they occur, are closely-spaced and can be mined together or on one quartz vein or the other according to the occurrence of gold in them.

Plate 13A shows the outlines of the stopes mined during the initial period of higher-grade production – to 1969. The better-grade ore occurs as shoots which trend directly down-dip; they are identified by the letters A, B ... G from north to south. Their shoot-like character is more evident in *Plate 13B* which is based on gold-contouring done by Bob Hewitt as part of a detailed study of the reef pattern and of the ore-bodies carried by it.[19] These contours show some tendency for the best ore to favour the upper parts of the various shoots. There is no stoping of any significance in the position where the Crown reef would intersect the downward projection of the Mararoa "favourable bed", i.e., below No. 22 level at 2500mN and below No. 29 level at 1006mN – see these and other cross sections in *Figure 55*.

19. R. Hewitt, Studies of Quartz Width, Gold Distribution and Structures in the Crown Reef, Report K/1714, 11/7/1969.

The northern limit of ore is B fault which is a zone of shearing about 35m wide. Mostly the individual shears are weak but, scattered through the zone, are impersistent fractures of greater strength, tending generally to be on the hanging-wall or the foot-wall of the zone. The fault zone trends north-westerly in plan view and dips north-easterly at about 60° or slightly less. Its attitude is similar to that of the Princess Royal fault at North Royal but it is a weaker feature.

Passage through B fault had a marked influence on the Crown reef fracture, as is evident in *Plate13*; these effects are more pronounced on No. 14 and No. 16 levels than on No. 19 and No. 22 levels. To the south of the fault, the Crown reef has a north-easterly strike and a flat dip. With-in and near the fault zone, the reef channel is disturbed and the overall effect is a dextral (or right-handed) off-set to the course of the reef. The central segments in these disturbed areas dip vertically. Similar but smaller features occur within or near ore-shoots C and E and, as with B fault and its disturbed area, there is a strong influence on ore occurrence.

Cross sections of these features are given in *Figure 56A* – at the positions marked in *Plate 13A*. They are structures of considerable strength. For instance, the central component in section e-e is 30m high, the height of a ten-storey building; the stoping on it is 3m or so wide, equal to the width of some sitting-rooms. Off-sets on the other cross sections are similar but slightly smaller. Cross sections a-a and d-d show vein-loops, with the added feature in section d-d of fold-like "wriggles" in the pattern of the quartz reefs.

The central components of these structures are complex. It is common to have two or more vertical shears, each carrying laminated quartz, and for these to be inter-connected by spurry quartz, i.e., on tension fractures. Fold-like crumpling is more common than could be shown in small-scale cross sections. Other indications of complexity, including vein-loops, are seen in the cross sections at 2301mN and 2500mN in *Figure 54* – in both cases, where the reef fracture passes through B fault.

Over the years, attempts have been made to explain these structures but none of them has proved satisfactory. More recently, it has been decided to describe their central segments as "verticals", thus emphasising their predominant attitude without ascribing an origin. The nearest relatives of the verticals may be the cross reefs seen elsewhere in the district; they strike E-W and dip very steeply but have in places a tendency to trend N.W. – see Cumberland area, *Plate 18C*.

In his study of the mine, Bob Hewitt made a detailed survey of the stopes above and below No. 19 level to trace changes in attitude of the vertical that is central to the occurrence of ore-shoot C. His results are shown in plan projection in *Figure 56B* and in cross section in *Figure 56C*.

At No. 19 level, the central component is vertical – cross section A – A in *Figure 56C* and cross section b – b in *Figure 56A*. Below No. 19 level, the dip remains vertical, as is indicated by the hatchured zone which traces the intersection of the central segment with normal reef above and to the north and with its companion below and to the south. Above No. 19 level, this vertical reef rolls to a flat dip (section B – B) and, at higher positions, as is indicated by the widening of the hatchured area, it becomes flatly east-dipping merging, in fact, into the normal Crown reef shear. Hewitt's conclusion was that the vertical is an integral part of the fracture pattern and not a feature super-imposed on it. This view is supported by a study (mainly petrological) done by Eric Cameron.[20] He

20. *E. Cameron, The Dyke Pattern at Norseman and its Relationship to Ore on Crown Reef, Report K/1685, 1/2/1968.*

noted, where No. 14 level north enters B fault, that laminated quartz on normal flatly-dipping reef rolls to a vertical attitude, sharply and without rupture.

The verticals and their neighbouring segments of normal reef form Z structures in plan view. They are different from the S structures of the other reefs as their central components, the verticals, relate partly or wholly to shearing and not to tension fracturing. The central component trends north-westerly, not north-easterly, and its vertical attitude contrasts with the south-easterly dip inherent in the development of an S structure *(Figure 17)*. In an S structure, a north-westerly trend is local to its steeply-dipping central core; in a Z structure, the north-westerly trend is through-going. In consequence, although both structures are central to ore occurrence, a Z structure is seen more clearly and is reflected in a strongly north-westerly trend in the ore pattern.

The diamond drilling used in the exploration for the Crown reef disclosed a stratiform unit, the Blue Bird gabbro, which had been injected into a zone containing several slate bands prior to the folding of the district. It is the first such feature to be found at Norseman in a position close to ore occurrence. The base of the Blue Bird gabbro was traced in the underground workings but with difficulty, due to wall rock alteration along the reef channel and to the paucity of useful cross-cuts. The general course of its intersection with the Crown reef is shown in plan projection in *Plates 13, 14 and 15C*. The base of the Blue Bird gabbro is shown in cross section in *Figures 54 and 55*.

In Plates 13, 14 and 15C, the upper limit of ore is A fault. This is a weak feature and its influence stems from the probability that it follows the upper limit of the Blue Bird gabbro. Above the Blue Bird gabbro, surface diamond drilling has shown the Crown reef channel to be weak and dispersed with a patchy development of quartz which is uniformly low grade.

In Plate 13B, high grade ore shows some predilection for the Blue Bird gabbro. Horizontal continuity declines at no great distance below and ore occurrence is reduced to fingers extending downwards into the underlying basalts – trends determined by the Z structures and the course of B fault.

Mining

Stoping commenced on the Crown reef during 1957 and was soon built up to a level comparable with that of the North Royal mine, the other main producer. At the time, the price of gold was low and operating costs were rising steadily. In consequence, mining on the Crown reef was confined to the better grade areas shown in *Plates 13A and 13B*.

However, by 1969, the ore supply at Princess Royal was dwindling and mining on the Crown reef had to be expanded to meet the short-fall. Stoping was extended laterally between the higher grade shoots and in depth below and between them. Some stope pillars were mined and the higher grade areas were tested, often successfully, for parallel veins that had been missed or left un-mined in earlier years. This extension of mining is shown in *Plate 14* and its effect is evident in the figures given below for production from the Crown reef.

Years	Tonnes	Average Grade (g/t)
To 1969	1,249,000	15.1
1969 – 1988	1,768,000	8.2

THE BLUE BIRD SEQUENCE

The surface diamond drilling for the Crown reef and some of the drilling in the Princess Royal area disclosed identifiable strata in close proximity to the orebodies – for the first time at Norseman. They are shown in cross section in *Figure 57A*. Cross sections 610 mN, 1260mN and 2955mN are in the Crown area and cross sections 635mS and 00 are at Princess Royal, as is shown in *Figure 57B*.

The sequence in the Crown area is the Blue Bird gabbro, overlain by the Royal amphibolite which is overlain in turn by the Crown basalt. These are sills intruded prior to folding into a zone within the basaltic lavas of the Norseman area which contains numerous lenses of slate. Two of the more persistent slate lenses are the Empress slate and the Gee Cee slate lying respectively above and below the Blue Bird gabbro. Only the base of the Blue Bird gabbro can be recognised at Princess Royal. It is seen at 635 m S in drill hole PRS 64, at scattered positions along the western sides of the open cuts and at depth in the North Royal shaft at coordinate 00.

The only indication of movement on B fault is given by the cross section at 2955 mN, *Figure 57A*. This interpretation shows the maximum possible movement.

The Blue Bird sequence was studied by Cor Bekker [21] who extended his work to a re-study of the whole Norseman area. His main sources of data were the cores of the surface diamond drilling traverses, underground mapping and some surface observation. He sub-divided the rocks of the area into the stratigraphic succession given below:

Desirable Pillow Lavas

Crown Basalts }
Royal Amphibolite } a composite sill
Blue Bird Gabbro }

Mararoa Pillow Lavas

Agnes Venture Slates

Kingswood Basalts

Noganyer Group Sediments

In the Crown area, the best stratigraphic marker is the base of the Blue Bird gabbro. It is coarser in grain-size than the underlying Mararoa basalts and shows large aggregations of feldspar crystals in many areas. As these reach a size of 25cms, they can give a spectacular rock, easily recognised. However, this feature is intermittent and, where it is absent, a study of the full succession may be needed to give identification of the Blue Bird sequence and its limits of occurrence. A complication is the occurrence of east-dipping gabbro dykes, particularly in the main productive area of the Crown mine. These dykes are younger than the westerly-dipping Blue Bird gabbro but share many of its features of texture and mineralogy.

Figure 58A shows the interpreted surface position of the Blue Bird gabbro in the Crown area and the relationships to it of its associated orebodies. This figure is partly schematic. The near-surface orebodies of the Golden Crown and Blue Bird mines are shown in their true positions, as is the Northern Star orebody at the northern end of the

21. *Cor Bekker, The "Greenstones" of the Norseman District, Report K/1330, 13/12/1962.*

Norseman reef. However, to assist presentation, the Crown orebody has been placed in a position in relation to the Blue Bird gabbro and B fault equivalent to that occupied by it in the underground workings – see *Plate 14.* A similar adjustment has been made for the small stope on the Mararoa reef above No. 11 level at the Regent shaft.

The orebodies on the Blue Bird reef lie in the Blue Bird gabbro and the small stope on the Mararoa reef in the basaltic rocks immediately below it. The upper part of the Crown relates to the Blue Bird gabbro but ore occurrence extends some distance below it into the underlying Mararoa basalts. The Northern Star orebody lies in these basalts towards the lower limit of the Crown stratigraphic position; it may be the remnant of a bigger orebody now largely removed by erosion.

The position as Princess Royal is shown in *Figure 58B.* The base of the Blue Bird gabbro is known over the length of the North Royal orebody but mostly as a general trend; the best specific point is at 635m S in PRS64 – *Figure 57A.*

The position at the old Princess Royal mine is an interpretation. It is based on a deep-lead to the west of the mine, the probability that its gold came from the erosion of an upward prolongation of the known orebody and, in consequence, that the base of the Blue Bird gabbro lies to the west of the present reef outcrop. The position shown in *Figure 58B* requires an off-set – due either to a sharp bend in strike or to faulting. A fault seems the more likely as it would align with a zone of shearing mapped by Cor Bekker on the No. 7 and No. 9 levels of the mine. This shearing strikes north-westerly and dips steeply north-easterly.

At North Royal, ore extends upwards into the Blue Bird gabbro but occurs mainly in the underlying basaltic rocks, i.e., the favourable part of sequence is similar to that of the Crown orebody. At the old Princess Royal mine, a similar relationship is likely.

To the north of the Princess Royal fault, the Blue Bird gabbro was intersected by the extensive pattern of surface diamond drilling shown in *Figures 43 and 44.* They give the surface position shown in *Figure 58B.* The trend is north-easterly and is shown to become increasingly so by an outcrop on the shore of Lake Cowan and another some distance beyond it on an island in the lake.

In both the Crown and North Royal mines, ore occurrence drifts downwards at the southern ends of stoping below the base of the Blue Bird gabbro. This indicates the weakness in using stoping outlines to define "favourable beds", as had to be done in earlier days. Princess Royal *Plates 11 and 12* are in error to some degree therefore. The favourable zone in *Figure 11A* is correct but the favourable bed, of which it forms a part, is wider particularly up-dip. The notional folds of *Plates 12A and 12B* should be smaller in size therefore. However, strong disturbance by the Princess Royal fault is evident in Figure 58B and even more so in *Figure 59* which relates the Blue Bird gabbro to the general geology. It is more probable than not that this disturbance included significant folding

It is clear in *Figure 59* that the trend of the Blue Bird gabbro relates to that of the western margin of the Noganyer Group. In the productive area at Norseman, it follows a simple course. A minor off-set is caused by B fault and a disturbance of similar size is possible to the north of the old Princess Royal mine. The only real interruption to the general trend is caused by the Princess Royal fault.

Figure 59 shows the Blue Bird gabbro over a length of about thirteen kilometres but, of this length, only 4 kms have been productive, slightly less than 2 km at the Crown and slightly more at Princess Royal. Over the years, the intervening ground and that to the south of the Crown and to the north of Princess Royal have been tested by extensive traverses of surface diamond drilling but without result. The intersection of a quartz reef with a favourable stratigraphic zone is not sufficient; additional factors are needed to yield an ore occurrence.

EAST-DIPPING DYKES

In his study of the Norseman area, Cor Bekker found that the westerly-dipping volcanic sequence of the Crown area is cut by many east-dipping dykes. These dykes are mainly gabbroic and tend generally to be similar in texture and mineralogy to the gabbro of the Blue Bird sill. An east-dipping dyke was identified at Princess Royal, as is shown in the cross-sections at 635 m S and at 00, *Figure 57A*. Bekker found that this dyke is coincident with the Princess Royal reef throughout the length of the two mines; it has been recrystallised to actinolite – tremolite – chlorite schist and is of course the medium to coarse-grained actinolite schist of earlier years.

East-dipping gabbro dykes are numerous in the Crown mine area and were recorded in generalised interpretations by Cor Bekker. Some years later, they were mapped and studied in detail by Eric Cameron.[20] The main elements are shown in *Plate 15*.

Plate 15A is a plan of No. 22 level showing a succession of dykes, marked A to H, which trend north-westerly and dip north-easterly at angles ranging mainly from 40° to 60°. Approaching the Mararoa reef, they tend to a more northerly strike. Dykes D and E lead into a large mass of gabbro which is described by Cameron as the Crown Main Dyke. It is a dome-shaped body, as is evident in cross-section in *Plate 15B*, with a flat base dipping gently south – *Plate 15A*.

The greatest dimension of the Crown Main Dyke is in the plane of the reef, as is shown in plan projection in *Plate 15C*. Its over-all trend is south-easterly and it terminates against B fault at its northern end. The intersections with the Crown reef of dykes C and D also trend south-easterly, as do the probable intersections of dykes A and B.

Ore occurrence is plotted as general outlines in *Plate 15C*. It is evident that ore- shoots A and B relate to the Crown Main Dyke and that ore-shoots C and D relate to dykes D and C respectively. Dykes A and B seem to have had no influence on ore-shoots F and G and ore-shoot E is not dyke-related.

It is likely that the fractures occupied by the east-dipping dykes are related to the stresses causing the zone of shearing described as B fault and so it is not surprising that the attitudes of the dykes are reflected in the ore-shoot pattern. The other main factor is the verticals of *Plate 13A*. The dykes and the verticals have similar trends in plan view but the dykes and B fault dip north-easterly and the verticals are transgressive to them – *Figure 56A*. The only common ground is that ore-shoot C follows both a dyke and a vertical and that the verticals within B fault mark the northern limits of both ore occurrence and the Crown main dyke.

The Regent shaft has a series of cross-cuts, which allowed Eric Cameron to trace dyke H from No. 25 level to the surface and to show that the Mararoa reef relates directly to it. However, as his interest was the Crown mine, he did not map the dyke along the drives of the Mararoa reef workings. It is now known that dyke H extends along the full length of the Mararoa reef, from the No. 4 orebody at the northern end to the No. 1 or Butterfly orebody at the southern end. [22] It has been named the Butterfly dyke. It is thought to be a composite dyke containing gabbroic and ultramafic intrusives, recrystallised to actinolite – tremolite – chlorite schist. It was regarded in earlier years as wall rock altered to fine to medium-grained actinolite schist.

The wall-rocks of the Crown are altered to actinolite-tremolite-chlorite schist but this zone is narrow. It is regarded in present thinking as an ultramafic dyke, now recrystallised. It was thought by Eric Cameron to be wall-rock alteration caused by the mineralising fluids.

20. *E. Cameron, The Dyke Pattern at Norseman and its Relationship to Ore on Crown Reef, Report K/1685, 1/2/1968.*

22. *Arthur Thomas, Chief Geologist, Central Norseman Gold Corp., 1983-87, Personal Communication.*

ORE PATTERNS

It has been known for many years that the reef pattern at Norseman has a north-westerly trend. It was recognised by Haddon King and was sketched in *Figure 8* of his report K/555. His sketch is reproduced as *Plate 16A*. It was based on the pattern of reef distrubution and ore occurrence and was regarded by Haddon "as a factor that should not be ignored in planning exploration".

The occurrence of ore at Norseman has proved to be more restricted than may have been envisaged by Haddon King. *Plate 16B* is a composite plan projection showing the stoped areas on the Crown, Blue Bird, Mararoa and Norseman reefs. Ore occurrence is confined to an area $3^1/_2$ km long and 1 km wide, with little indication of any particular trend at the northern end. The southern end of ore shows a north-westerly trend from the Crown to the Blue Bird to the Mararoa and possibly to the Norseman reef.

A north-westerly influence is seen of course on the Crown reef where B fault marks the northern end of ore. This fault was mapped by Eric Cameron on the No. 22 level of the Mararoa workings – see *Plate 15A*. It lies between the No. 3 and No. 4 Mararoa orebodies but seems to have had little effect on them. B fault is found in the upper levels of the Regent shaft where a small orebody on the Mararoa reef has been mined on its northern side. To the north of the Regent shaft, the Mararoa reef turns north-westerly, see *Figure 19*, suggesting a fault influence to Eric Cameron. This suggestion has been adopted and is shown as D fault in *Plate 16.*

The faults and the reefs are shown in cross section in *Plate 16D*. If the surface position of B fault is projected south-easterly, it aligns with a north-westerly trend in the Norseman reef to the north of the small Royal Dane mine and is seemingly the cause of this trend, *Plate 16*C. D fault is assumed to be parallel to B fault in *Plate 16D*; if its surface position is projected south-easterly, it fits into a break in the surface outcrop of the Norseman reef and seems to be the cause of it, whilst to the north of the Northern Star mine, the return to a north-westerly strike may relate to the Mt Barker fault. Near the southern end of the reef, F fault can be inferred in an area where the reef channel has never been found.

The Norseman reef presents, in an exaggerated form, a feature of the Mararoa reef – west-of-north strikes have steeper dips and east-of-north strikes have flatter dips. On the Mararoa reef, the latter are favourable for ore occurrence. This applies also at the southern end of the Norseman reef; however, in its northern half, where east-of-north strikes and flat dips are well-developed, there is an abundance of quartz but a paucity of gold. This led Des O'Driscoll to the disgruntled conclusion that perhaps the structures were too favourable and so were sealed-off by early quartz, non-auriferous.

The stoped areas on the Crown, Mararoa and Norseman reefs are shown in *Plate 17A*. They have been separated in the east-west direction to avoid the over-laps of *Plate 16B* and have been moved in the north-south direction to give alignment along the general direction of B fault. On each reef, the zones more favourable for ore occurrence have been marked. On the Mararoa reef, the folding hypothesis of *Plate 6A* has been abandoned as the course of the Blue Bird gabbro *(Figure 58A)* shows that a fold off-set is not possible. Instead, a favourable zone has been drawn using only a modest displacement on B fault; this has required the exclusion of a small but high-grade stope which is used as the northern limit of the No. 3 orebody in *Plates 4, 5 and 6.* The Mararoa favourable zone has been extended southwards to include the No. 5 orebody.

In relation to B fault, the No. 2 and No. 3 orebodies on the Mararoa reef correspond in position to the Crown orebody, as do the ore occurrences on the Norseman reef except, of course, that there is no significant ore between F and B faults. The No. 1 orebody on the Mararoa reef is partly or wholly to the south of the Crown position and the No. 5 orebody is farther south. The No. 4 orebody is to the north of B fault and the Northern Star orebody is farther north. In effect, the north-westerly trending faults are "stepping stones" and the limit of ore occurrence moves northwards going eastwards from reef to reef.

The three favourable zones are brought to their correct relative positions in *Plate 17B*. Something is amiss as the northern end of the Mararoa zone is clearly at variance with the Crown favourable zone. As the Crown orebodies relate to a conformable feature, the Blue Bird gabbro, the Mararoa favourable zone, as drawn, cannot be wholly stratigraphic. An explanation is attempted in *Plates 18 and 19.*

Plate 18A shows the five Mararoa orebodies as general outlines and relates them to a favourable zone conformable with stratigraphy. Its basis and its projection to the surface *(Plate 18B)* are given in the cross sections shown in *Plate 19.* In the 1753mN cross section, the base of the Crown favourable zone (the Blue Bird basalts) projects up the dip to the top of the northern end of the Mararoa No. 3 orebody.

The next step is hypothesis and is shown in *Plate 19B.* The underlying assumption is that the Crown favourable zone is stratigraphic and that, going south, ore occurrence contracts away from its limits. This stratigraphic zone is shown in *Plate 19B* as the Blue Bird basalts. Their base, when projected up-dip in cross sections 1067mN and 335mN, conforms with the upper limit of the No. 2 orebody, *Plate 18A*; from there, it is logical to extend southwards along the top of the No. 1 orebody. From 1067mN to the southern end of the Butterfly mine, this position is marked by a sharp change from quartz and high-grade ore to weaker shearing, erratic quartz and no ore.

The key cross sections at the southern end are at 250mS and 830mS, through the Mildura and Viking mines respectively. The base of the favourable zone on the Norseman reef has been extended below the position shown in *Plate 17A* to include stoping done in recent years below the Old Viking and Mildura mines. The revised outline is shown in *Plate 19C* where it is called the Viking favourable zone. The Norseman reef outcrop, assumed to be the top of this zone, aligns on the two cross sections with the base of the Mararoa favourable zone and so the Viking zone passes below the bottom of the Butterfly mine.

Plate 18A shows the Mararoa No. 5 orebody in the southward prolongation of the Viking zone and not in the Mararoa zone as was shown in *Plate 17A.* This change gives sympathy with the stratigraphic trends. At and north of the Regent shaft, the top of the Mararoa favourable zone has been extended parallel to the trend of the Blue Bird gabbro; no offset is shown at D fault as the movement along it, if any, is not known. This interpretation implies that the No.4 orebody has migrated upwards into the Blue Bird basalts under the influence of D fault. The small orebody to the north of B fault in the upper reaches of the Regent shaft lies near the top of the Blue Bird basalts.

Plate 18B shows the Mararoa and Viking favourable zones projected to the surface and their relationships to the stratigraphic trends given by the Blue Bird gabbro and the

Agnes Venture slates. The combined favourable zone (Mararoa plus Viking) has a general sympathy with the two stratigraphic features but there is a divergence between its top and the base of the Blue Bird gabbro at the southern end.

Plate 18B shows also two lines marked M9 and M13, respectively. They relate to a study of the volcanic sequence, initiated in 1982 by Pat MacGeehan[23], in which the Mararoa basalts were sub-divided into the individual lava flows. This was a difficult task as the basalts are recrystallised and so the individual flows had to be identified in their modified forms. This was handled successfully, allowing the recognition of a succession of flows from M1 above the Agnes Venture slates to M25 at the base of the Blue Bird gabbro.

The surface was mapped by Richard Keele[24] over an area extending from the environs of the Phoenix shaft to the southern limit of outcrop. M9 in Plate 18B is the top of lava unit M 9, the most continuous member shown in his map of the area. M13 is the top of unit M13 and has been taken from a surface map prepared by Arthur Thomas[22] using the upward projection of data from surface diamond drill holes.

Various comparisons can be made in *Plate 18B.* In the main, the trends given by M9 and M13 support the view that the favourable zones are stratigraphic but, as is to be expected, with some disparities in detail. The crux is the thickening, going south, of the Blue Bird basalts as this is critical to the interpretation of the upper limit of the Butterfly orebody. The broader view given in *Plate 18C* suggests that this thickening, if actual, is no more than local.

The favourable bed hypothesis seems basically unchanged therefore – southerly – pitching orebodies within a favourable zone pitching gently north. However favourable zones do not extend from one reef to another. The Blue Bird basalts are unfavourable on the Mararoa reef except for a minor occurrence at its northern end. Equally, the Crown orebodies do not extend below the base of these basalts, except for an extension of ore-shoot E along a strong vertical and some low grade stoping to its north above and below No. 22 level, *Plate 14.* The favourable beds on one reef seem unfavourable on the other.

Incompatability is evident also in the Viking favourable zone. It carries the orebodies at the southern end of the Norseman reef, *Figure 45A.* It seems to be the host of the Mararoa No.5 orebody but, otherwise, is totally unfavourable on the Mararoa reef, even though it seems to be part of the Mararoa favourable zone. In section 850 N of ***Figure 14,*** the bottom of the No.2 Mararoa orebody is marked by a change to a very steep dip – at No.16 level where it passes into the Viking favourable zone. A similar change is seen in the cross sections at 250 mS and 830 mS, *Plate 19* This change to a steep dip slices like a knife through the Mararoa shear-link pattern from the environs of the Phoenix shaft to the southern end of the Butterfly mine and was fatal to ore occurrence. Conversely, the Mararoa zone is unfavourable on the Norseman reef at and north of the Phoenix shaft. As the choice of host seems to have been specific, it is not surprising therefore that the deep testing at Princess Royal did not find ore in the presumed position of the Mararoa favourable bed.

To the south of Norseman, the favourable zones found expression in two groups of cross reefs – the OK in the Blue Bird zone and the Cumberland group in the Mararoa zone *(Plate 18C).*

23. P. J. MacGeehan, Consulting Geologist – Minerals Exploration, W.M.C.

24. R.A. Keele, Emplacement and Deformation of Gold-bearing Quartz-veins, Norseman, W.A., Ph. D. Thesis, Univ. of Leeds, 1984.

22. Arthur Thomas, Chief Geologist, Central Norseman Gold Corp., 1983-87, Personal Communication.

The occurrence of ore-bodies and mines within the favourable zones is given below:

Favourable Zone	Orebodies and Mines
Blue Bird	OK, Blue Bird, Crown, Northern Star, Old Princess Royal, Princess Royal North
Mararoa	Cumberland, Mararoa No.1, 2, 3, 4 orebodies
Viking	Mararoa No.5, Viking, Old Viking, Mildura

Productions from those zones are:

Favourable Zone	Tonnes	Ounces	g/t
Blue Bird	6,076,000	2,846,000	14.6
Mararoa	3,174,000	1,147,000	11.2
Viking	493,000	181,000	11.4

The Blue Bird zone is the main producer of the area by all measures – the spread of ore along it, the number of productive positions, the tonnage mined and its average grade.

The reasons for favoured locales within these zones are unclear except perhaps at Princess Royal where the Princess Royal fault is an obvious factor. It aligns with the position where the Noganyer sediments change from a northerly to a north-easterly trend *(Figure 59)* and its east-block-south movement is compatible with this change. The North Royal orebody relates to movement along the Princess Royal fault and the old Princess Royal orebody and its fault are probably satellite to it.

The position is different in the Norseman area as its group of orebodies lies opposite the undisturbed flank of the Noganyer sediments. B fault, although significant to ore occurrence on the Crown reef, is a minor feature in regional terms. The succession of faults along the Norseman reef looks impressive *(Plate 17A)* but none causes any real offset of the Agnes Venture slates. Rather than faults, they are more likely to be northwesterly trending segments of the reef fracture. However, D fault does seem to be important on the Mararoa reef. It marks the northern end of ore occurrence and the change to a west-of-north strike and a steep dip which extends as far north as the norite dyke. [(25)] Extensive surface diamond drilling of this area has found some reef but no ore.

Within the ore areas, many north-westerly trends are evident – the Z structures, B fault and the north-westerly trending dykes on the Crown reef, the S structures and the north-westerly trending orebodies on the Mararoa and Norseman reefs and the various north-westerly trending faults on the three reefs. Although no reasons can be ascribed, it is clear that there is a favoured area at Norseman and that, within it, north-westerly influences are important.

The Mararoa, Norseman and Princess Royal reefs trend north-south and all show flatly-pitching shear-link structures. At Princess Royal, this pattern may have been modified by disturbance relating to the Princess Royal fault, at least in the open-cut area and possibly in the underground part of the mine – link 6, link 9 at 300mS in *Figure 37A* and link 5, link 9 at 400mS in *Figure 37B.*

25. *Arthur Thomas, Review of Drilling Targets, Memo to Resident Manager, 7/11/1984; up-dated, June 1987.*

The Crown reef trends northeasterly – southwesterly, except where it joins the Mararoa reef at its northern end, and has a uniformly flat dip within the whole of the ore-bearing area. Its Z structures are unique to it and have an origin different from that of the S structures of the other reefs. The reef has been followed on No.22 level to the south of the area shown in *Plate 13* (see *Figure 60)* and is known from surface diamond drilling to maintain its course for some distance farther south. It is a persistent fracture. The Crown has produced more ore than either the Mararoa or Princess Royal reefs and so the channel followed by it must be an important, if unexplained, component of the fracture pattern of the Norseman district.

PRODUCTION HISTORY

The production history of Central Norseman from 1937 to 1988 is shown graphically in Plate 20. During these years the company treated 8,409,000 tonnes for the recovery of 3,452,000 ounces of gold, an average of 12.8 g/t.

The main contributors are listed below:

Reef	Tonnes	Ounces	g/t
Mararoa	2,314,000	745,000	10.0
Princess Royal	2,764,000	1,583,000	17.8
Crown	3,042,000	1,050,000	10.7
Totals	8,120,000	3,378,000	12.9

They have produced 97 per cent of the ore mined by Central Norseman and 98 per cent of the gold recovered.

The remaining production came from the Lady Miller during the 1940's and, in recent years, from the OK and from a recent discovery made by Central Norseman, the Scotia deposit about 40 km south of Norseman. The OK is a cross-reef relating to the Blue Bird gabbro and is shown in *Plate 18C.* The Lady Miller lies in banded iron formation within the Noganyer Group about 8kms to the south of Norseman.

Production from the Mararoa, Princess Royal and Crown reefs over-lapped neatly, as is evident in *Plate 20,* and allowed Central Norseman to maintain an improving rate of production from 1937 into the late 1960's. There was a significant increase in grade in 1953 when the new discoveries at Princess Royal came fully into production, supported after 1958 by the mining of the high-grade ore of the individual ore-shoots on the Crown reef. Grade decreased after 1965 as production fell away at Princess Royal and mining on the Crown reef was extended into the lower grade areas.

There was a significant improvement in 1976 when the near-surface ore at Princess Royal came into production, initially by under-ground mining followed a year or two later by the development of the open-cuts. In recent years, grade has declined and has been countered by increasing production – a step made possible by the building of a new carbon-in-leach treatment plant.

The figures given below are, in themselves, a commentary on the nature of the field:

Outcropping Orebodies	Tonnes	Ounces	g/t
Princess Royal	183,000	152,000	25.8
Norseman Reef	102,000	53,000	16.1
Cumberland	68,000	65,000	29.9
O.K. Reef	11,000	15,000	43.3
Other Reefs	113,000	67,000	18.5
Totals	478,000	352,000	22.9

Continuation of table on page 61.

Concealed Orebodies	Tonnes	Ounces	g/t
Old Mararoa Co.	352,000	172,000	15.2
Butterfly	700,000	225,000	10.0
Viking	57,000	54,000	29.5
Central Norseman	8,120,000	3,378,000	12.9
Totals	9,229,000	3,829,000	12.9
Grand Totals	9,707,000	4,181,000	13.4

Notes:

1. The old Mararoa and Viking orebodies had weak outcrops covered by soil and hill-wash. They were not found and mined until the early 1900's.
2. Norseman Gold Mines treated a considerable tonnage from the IronKing mine, a banded iron formation orebody. The figure for the Butterfly mine is estimated.
3. The figures for Central Norseman include only the Mararoa, Princess Royal and Crown reefs.

The concealed orebodies account for 95 per cent of the ore mined from the main producers of the field and have yielded 91.5 per cent of the gold recovered. The Crown orebody is geologically concealed – its top is near No. 14 level, 270 m vertically below the surface. The North Royal is different as its reef outcropped at the Archaean surface and its concealment is due to covering by the relatively recent Miocene sediments. The Mararoa reef is the most striking. Ore has now been mined over a length of 4km yet, for all practical purposes, none of it outcropped despite the fact that the top of ore is little more than 50m from the surface in the Butterfly mine.

Except for the No. 4 orebody, all of the ore on the Mararoa reef lies above the No. 22 level, i.e., at less than 500 m vertically below the surface. The deepest ore on the No. 4 orebody is at No. 29 level, 600 m below the surface. The effective bottom of the Crown orebody is No. 25 level, at a depth of 510m, with only minor extensions below it. The deepest mining at Princess Royal is at No. 10 level, less than 300m vertically below the surface.

Gold produced and dividends paid are shown graphically in *Figure 61.*

Gold production took a step upwards in 1953 when the North Royal came into production, reaching 100,000 ounces per year and sustaining this level until 1966. The Crown reef became a major contributor in the later of these years. Ounces recovered then fell away as production from Princess Royal declined and the grade mined on the Crown reef became lower. The influence of the near-surface ore at Princess Royal is evident in the increase in gold produced in 1976 and its continuance at a high level in the following years.

The dividends paid are shown in *Figure 61B* and, in the main, reflect the amount of gold produced. The lean years were 1933 to 1942 and 1972 to 1976.

As money values have changed so much over the years, *Figure 61B* goes off-scale in the years after 1977. The vertical scale has been reduced in *Figure 61C* but, even here, the year 1986 goes off-scale. In this figure, the dividends of earlier years seem minuscule in relation to those paid since 1976. However, this is not so.

Figure 61D shows the dividends paid in terms of 1970 dollars. The basis is the award rate for engineering fitters in relation to the gold price of the day. The fitter's award rate was $21 per week in 1942, $65 in 1970 and is now more than $300 per week.

Figure 61D shows that the dividends paid before 1953 were not as low as they might seem – they averaged more than $500,000 per year in 1970 dollars. Increasing dividends from 1953 to 1966 become a steady rate of return in a period of declining money values. The high dividends of the last ten years or so reflect the change to a more realistic price for an ounce of gold.

The Central Norseman gave ample rewards, particularly to its major shareholder and supporter Western Mining Corporation Ltd.

IN RETROSPECT

Success at Norseman came from the pursuit of excellence. This is possible only if there is adequate and sustained support – financial, managerial and technical. The financial support came from W.S. Robinson, the managerial support from Lindesay Clark and the technical expertise from Terence Conolly.

W.S. Robinson is well-known, of course, to mining people in Australia. He dealt with people, at all levels, pleasantly, politely and courteously. This was not an "image", as is so common now-a-days; it was genuine. He was a kindly man, always helpful and generous, proud of his achievements during a long career of widely-ranging activity, and rightly so.

Before, during and after the First World War, W.S. Robinson became well-known and greatly respected in financial and mining circles in London and the U.S. He led the way in the foundation of Western Mining Corporation and organised a group to finance it. When things went badly early in the career of W.M.C., he was able to arrange further finance and to continue to do so for as long as needed. Without him, W.M.C. could never have survived nor could Central Norseman which was a major cause of money problems.

The technical support came from Lindesay Clark, a quiet and basically shy man but tough and determined never-the-less. He had imagination and vision and was always willing to follow-through on a project if he felt that it had potential. Central Norseman was one of his many problems and he supported every prospecting proposal at Norseman despite the long succession of disappointments in the early years. He did not give up easily and his tenacity was a major factor in the long hard search for success at Norseman. Although not the largest, Central Norseman was in many ways his greatest achievement. It was a critical success at a critical time.

After the war, I went on many trips with Lindesay to various mines and prospects – on some occasions, just the two of us. We became very friendly and this friendship lasted until his death. A consequence was that my wife got to know and like Lindesay and his wife Jane, a quiet but charming woman. We were their guests on many occasions and they ours, particularly Lindesay when in Kalgoorlie. Our working relationship was just as friendly until the late 1950s when we began to differ on some matters of consequence. The personal friendship continued but the working relationship was impaired. I resigned in 1967 soon after the death of my wife and used a minor event as the reason, as I did not wish to offend Lindesay. I suspect that he guessed the true reason but, if so, gave no indication to me. It was typical of him that he insisted that my consulting arrangements with B.H. South and Alcoa should continue. We met over the subsequent years on every convenient occasion to have a friendly chat and, more often than not, lunch at the Australian Club. It was a pleasure to have known him and Jane so well for so long.

Terence Conolly was as Irish as his name. He had all the charm of the Irish and some of their faults – at his best he was witty, voluble and entertaining. When things were going badly, he became moody, uncommunicative and sometimes unpleasant. My association with him lasted some ten years or so. Mostly I liked and admired him but, at times, I found him rather trying; on balance, I have fond memories of him. Certainly, I learned a great deal from him.

Terence learned the Harvard methods of geological mapping and interpretation during his years in Canada and applied them of course when he went to Norseman. Subsequently, he spread them throughout the operations of Western Mining. The purpose was to understand the relationship between ore occurrence and geological structure and to project the structural controls into the "wide blue yonder" to predict where another ore position could lie. This would then be tested, mostly by surface diamond drilling; when the results came, re-interpretation and another set of drilling targets was the order of the day, and so on, and so on. Conolly's message was: be imaginative, be bold – timidity is the road to failure! Inevitably, this leads to some disasters – the trick is not to persist too long on a false trail and to learn from the failures. This method has been followed ever since in the operations of Western Mining and has led in recent years to the finding, under the guidance of Roy Woodall, of Kambalda and the Olympic Dam deposit at Roxby Downs.

At Norseman, Conolly mapped the mine levels aided by Jim Pocock as his "penciller". Alan Blatchford, another young geologist, and I surveyed and mapped cross sections at 50 ft. intervals throughout the length of the mine. Conolly was a hard task-master on both himself and those with him and we worked like beavers. At night, Terence transferred his mapping notes onto tracings of the mine level plans, Jim prepared the next day's mapping sheets, and Alan and I plotted our cross sections. This mapping and draughting job was done in about six weeks.

We stayed at the old Norseman pub, a somewhat derelict ruin of earlier days. The accommodation was poor and the food was worse; however, we were young (including Conolly), we had a job to do and enjoyed doing it.

The only large table at Norseman was in the old Shire office (also near-derelict), so Terence with his Irish charm talked his way into the free and unfettered use of it for a month or so to get a space large enough to do his interpretation. He worked away at this task and Alan and I did the donkey-work of transferring data points from plan to cross section and vice-versa, inking and colouring finished interpretations, etc. This was my training in shear-link interpretation – watching Terence as best I could to see what he was doing, why and how.

When the Norseman job was finished, Terence and Alan went to Kalgoorlie and I was given the task of mapping and interpreting the geology of the workings on the Norseman reef – essentially the Viking and Old Viking mines. Conolly wished to try a method of shear-link interpretation different from that used on the Mararoa reef. I was given about ten minutes or so of instruction on what to do and how to do it and that was it. It was a sink or swim affair; I managed a respectable "dog-paddle". This was followed during 1935 by a study of the old Princess Royal mine, already described in the text.

After the war, I kept in close touch with Norseman until about 1957-58. During this period, I made many mistakes as has been narrated. I regret none of them as they were an essential part of the learning process. If you don't try, you don't get new data and so cannot develop new and better hypotheses. After that, I was little more than a "visiting fireman" as Norseman was going well and I had many tasks elsewhere.

On the geological side, Haddon King made the most effective contribution with his favourable bed hypothesis. It has been used since 1941 to find all the orebodies mined subsequently at Norseman. His PRS 5 at Princess Royal led ultimately to the finding of the North Royal orebody. I mentioned to Haddon several years ago that, although it proved to be a bit of a fluke, it was in fact the "discovery hole" and so rated highly in the history of Central Norseman. Being a modest man, he was reluctant to accept this view but agreed ultimately that it was a significant event.

Terence Conolly had the misfortune to work on Norseman during its early and difficult years and many of his ideas were proved by events to be wrong. However, the greatest disaster, the Ajax Shaft project (partly his), disclosed the Agnes Venture slate and gave Haddon King the vital clue in the development of his favourable bed hypothesis. Despite the failures, Conolly made a great contribution during the critical years of the Central Norseman operation. From the start, he recognised the potential of the Norseman field and, when necessary, proclaimed it loudly and longly. Beyond this, W.S. Robinson had complete confidence in Conolly and this led him to extracting money when needed from some very "reluctant dragons" in London. Great credit goes to Conolly for the part he played in helping to sustain a company that so easily could have foundered.

Three-dimensional geological interpretation involves charting a correct course through many alternatives in key areas; it is far from easy and is time-consuming. The last few years have given me great pleasure therefore as I have been able to devote unlimited time to the interpretation of the vast amounts of data collected at Norseman over the years. This was not possible during my years as Chief Geologist of Western Mining Corporation, nor was it possible for the mine geologists who were busy with daily affairs. The most interesting and difficult part of my recent work was Princess Royal as it has a complex vein structure and there is a paucity of data in critical areas in the open-cuts.

The "unsung heroes" of Norseman geology are the many geologists, mostly young, who mapped the mine levels and the stopes, the latter leading to the preparation of cross sections. The task is laborious and never-ending and the picture builds-up so slowly that few of them ever see an over-all evaluation. My interpretations are built on their mapping and I thank them sincerely for their work.

I had great help from Arthur Thomas, Chief Geologist of Central Norseman until his retirement at the end of 1987, and then from his successor Kevin Johnson. They took a real interest in my work and helped by providing plans and other data when requested.

A key point in the Central Norseman Story, in fact the key point, is the illustrations. The Mararoa illustrations were done by Diana Kurauskas and modified subsequently by Vicki Jackson to conform to her style. Vicki draughted the remainder of the figures and plates – in other words, nearly all the illustrations. Vicki put her heart into the job and worked tirelessly in seeking improvement on improvement in all the diagrams. She was a really great help to me as is evident in the results of her work.

Unfortunately, Vicki had to leave at the end of October 1988 as her husband had been transferred to New Zealand. She was determined to complete the first draft of the remaining few diagrams, a couple of them complex plates. She succeeded, with only a few hours to spare, and was completely exhausted at the end of this great effort – and then had to pack to go to New Zealand a day or two later!

The preparation of a diagram is a long process, changing this, altering that, again and again. Greg Morris has completed Vicki's work. He checked all illustrations against the text and vice-versa and searched for minor errors. He found some that had eluded Vicki and me – we had become too familiar with what had been drawn. Greg's meticulous work has been a great help.

An un-related point. The Chief Geologists of Western Mining have been:

1934 – 35 Terence Conolly;

1936 – 42 Haddon King;

1945 – 67 Don Campbell;

1967 – Roy Woodall.

Between us, Roy and I have covered 44 years, with more to come from him – stuff for the Guinness Book of Records? It is certainly a measure of the stability of Western Mining Corporation and of support for its heads of departments.

Major discoveries and developments by Western Mining Corporation are listed:

1. Central Norseman.
2. Mt. Charlotte. It's potential as a low grade, low cost, high volume producer was recognised by Lindesay Clark and, after some surface diamond drilling, had the enthusiastic support of Terence Conolly. Both were bitterly disappointed when it was turned down by London in 1938. Lindesay stuck to his guns and got it into production in the late 1950s. In the late 1970s, it became very profitable and has remained so since then.
3. Alcoa of Australia. Following the finding of Weipa, Lindesay thought that W.M.C. should get into aluminium if possible and wanted to try the Kimberley area in W.A. He discussed it with me and I suggested the Darling Ranges, knowing that H.W.B. Talbot had looked at them in 1940 for Zinc Corporation. This led to testing of the area and the formation of Alcoa of Australia in 1961.
4. Kambalda. This came from the background knowledge of mafic-ultramafic geology of Roy Woodall and his recognition, on sketchy data, that the Kambalda dome could have a potential for nickel orebodies.
5. Roxby Downs. This was a team effort and was the result of good conceptual thinking by its key members.

The progression, in terms of size and value, is exponential – the Olympic Dam deposit at Roxby Downs is one of the biggest and best ore-bodies in the world. The time-span is 1938-39 at Norseman to 1975 at Roxby Downs, giving a "strike-rate" that is not high but is perhaps better than average.

These successes came from hard and imaginative thinking, backed by adequate support for the resulting projects. This approach reflected the attitudes of Lindesay Clark and, in the earlier years, of the people supporting him, W. S. Robinson and Sir Colin Fraser.

Such a policy has inherent risks and there were many crises during the early years of the Western Mining operation. However, the greatest problem arose during the 1950s when a major project was attempted in the Southern Cross area (Great Western Consolidated). It was a disaster, both technically and financially, entailing Western Mining in ever-increasing loans, ultimately worthless, to a company that had little chance of success. Western Mining was by then on its own and its survival is due, in the main, to the high-grade North Royal ore-body and the higher grade portions of the Crown Reef ore-bodies. Dividends and loans from Central Norseman were the key factor in keeping Western Mining "afloat" from the late 1950s until the late 1960s, when the Kambalda nickel operation gave welcome relief to a company that had been, at times, close to failure.

REFERENCES

1. W.D.Campbell, G.S.W.A., Bulletin No.21, 1906.
2. E. de C. Clarke, C. Teichert, J.R.H. McWhae, Tertiary Deposits near Norseman, W.A., Proc. Roy. Soc. of W.A., 1948.
3. H. Cloos, Zur Experimentellen Tektonik I and II, Die Naturwissenschaften, Vol. 18, No.34, 1930; Vol.19, No.11, 1931.
4. H.J.C. Conolly, The Phoenix Prospect, Norseman, W.A., Report 55 T/1, April 1934.
5. H.J.C. Conolly, A Contour Method of Revealing Vein Structures, Economic Geology, Vol.31, pp 259-271, 1936.
6. H.J.C. Conolly, The Norseman Prospect, Report K/53, 18/1/1935.
7. H.F. King, Geological Structures and Ore Occurrence at Norseman, West Australia, Report K/555, 24/4/1941.
8. J.D. Campbell, Norseman Development, Report K/631, 4/6/1946.
9. J.D. Campbell, Report on Princess Royal Prospect, 1935.
10. J.D. Campbell & D. O'Driscoll, Princess Royal Exploration, Report K/716, 19/11/1947.
11. J.F. Fradd, Chief Geologist, Central Norseman, 1971-1976, Personal Communication.
12. G.M.F. Hopkins, Chief Geologist, Operations (W.A.), Personal Communication.
13. J.D. Campbell, Princess Royal Exploration, Memo to F.F. Espie Snr., 7/8/1946.
14. J.D. Campbell, Crown Reef and Princess Royal Exploration, Report K/1096, 14/6/1957.
15. D. O'Driscoll, Geological Notes for the Period 22/6/1949 to 13/9/1949.
16. D. O'Driscoll, Norseman Reef Drilling Proposals, Report K/715, 14/11/1947.
17. D. O'Driscoll, Norseman Reef Drilling, Report K/767, 17/11/1948.
18. D. O'Driscoll, Proposed Blue Bird Drilling, Report K/856, 10/8/1950.
19. R. Hewitt, Studies of Quartz Width, Gold Distribution and Structures in the Crown Reef, Report K/1714, 11/7/1969.
20. E. Cameron, The Dyke Pattern at Norseman and its Relationship to Ore on Crown Reef, Report K/1685, 1/2/1968.
21. Cor Bekker, The "Greenstones" of the Norseman District, Report K/1330, 13/12/1962.
22. Arthur Thomas, Chief Geologist, Central Norseman Gold Corp., 1983-87, Personal Communication.
23. P.J. MacGeehan, Consulting Geologist – Minerals Exploration, W.M.C.
24. R.A. Keele, Emplacement and Deformation of Gold-bearing Quartz-veins, Norseman, W.A. Ph.D. Thesis, Univ. of Leeds, 1984.
25. Arthur Thomas, Review of Drilling Targets, Memo to Resident Manager, 7/11/1984, up-dated, June 1987.